Biosafety and Bioethics in Biotechnology

Multidisciplinary Applications and Advances in Biotechnology: Contributions from the Biotechnology Society of Nigeria Working Groups

Series Editors:

Benjamin Ewa Ubi

Department of Biotechnology / Biotechnology Research & Development Centre, Ebonyi State University Abakaliki, Nigeria

Abdulrazak B. Ibrahim

Department of Biochemistry, Ahmadu Bello University, Zaria, Nigeria / Forum for Agricultural Research in Africa (FARA), Accra, Ghana

Sylvia Uzochukwu

Biotechnology Centre, Federal University Oye Ekiti, Nigeria

Nwadiuto (Diuto) Esiobu

Microbial Biotechnology Laboratory, Florida Atlantic University, Boca Raton, FL., USA, / Applied Biotech International Nigeria Ltd., Wuye, Abuja, Nigeria

Arinze Stanley Okoli

GenØk Centre for Biosafety, Tromso, Norway

Christpeace Ezebuiro

Clean Energy Division, National Biotechnology Development Agency (NABDA), Abuja, Nigeria

Charles Oluwaseun Adetunji

Edo State University Uzairue, Nigeria

Fermentation and Algal Biotechnologies for the Food, Beverage and Other Bioproduct Industries

James Chukwuma Ogbonna, Sylvia Uzochukwu, Emeka Godfrey Nwoba, Charles Oluwaseun Adetunji, Nwadiuto (Diuto) Esiobu, Abdulrazak B. Ibrahim, and Benjamin Ewa Ubi

Agricultural Biotechnology, Biodiversity and Bioresources Conservation and Utilization

Olawole Obembe, Emmanuel Olufemi Ekundayo, Arinze Stanley Okoli, Abubakar Gidado, Charles Oluwaseun Adetunji, Abdulrazak B. Ibrahim, and Benjamin Ewa Ubi

Medical Biotechnology, Biopharmaceutics, Forensic Science and Bioinformatics
Hajia Mairo Inuwa, Ifeoma Maureen Ezeonu, Charles Oluwaseun Adetunji, Abubakar Gidado, Emmanuel Olufemi Ekundayo, Abdulrazak B. Ibrahim, and Benjamin Ewa Ubi

Biosafety and Bioethics in Biotechnology: Policy, Advocacy, and Capacity Building
Sylvia Uzochukwu, Arinze Stanley Okoli, Nwadiuto (Diuto) Esiobu, Emeka Godfrey Nwoba, Christpeace Nwagbo Ezebuiro, Charles Oluwaseun Adetunji, Abdulrazak B. Ibrahim, and Benjamin Ewa Ubi

Microbiomes and Emerging Applications
Nwadiuto (Diuto) Esiobu, James Ogbonna, Charles Oluwaseun Adetunji, Olawole Obembe, Ifeoma Maureen Ezeonu, Abdulrazak B. Ibrahim, and Benjamin Ewa Ubi

Bioenergy and Environmental Biotechnology for Sustainable Development
Akinola Rasheed Popoola, Emeka Godfrey Nwoba, James Chukwuma Ogbonna, Charles Oluwaseun Adetunji, Nwadiuto (Diuto) Esiobu, Abdulrazak Ibrahim, and Benjamin Ewa Ubi

For more information about this series, please visit: https://www.routledge.com/Multidisciplinary-Applications-and-Advances-in-Biotechnology/book-series/MAABNIG

Biosafety and Bioethics in Biotechnology

Policy, Advocacy, and Capacity Building

Edited by
Sylvia Uzochukwu,
Nwadiuto (Diuto) Esiobu,
Arinze Stanley Okoli,
Emeka Godfrey Nwoba,
Christpeace Nwagbo Ezebuiro,
Charles Oluwaseun Adetunji,
Abdulrazak B. Ibrahim, and
Benjamin Ewa Ubi

CRC Press
Taylor & Francis Group
Boca Raton London New York

CRC Press is an imprint of the
Taylor & Francis Group, an **informa** business

First edition published 2022
by CRC Press
6000 Broken Sound Parkway NW, Suite 300, Boca Raton, FL 33487-2742

and by CRC Press
4 Park Square, Milton Park, Abingdon, Oxon, OX14 4RN

CRC Press is an imprint of Taylor & Francis Group, LLC

Library of Congress Cataloging-in-Publication Data
Names: Ubi, Benjamin Ewa, 1968- editor. | Biotechnology Society of Nigeria, issuing body.
Title: Multidisciplinary applications and advances in biotechnology :
contributions from the Biotechnology Society of Nigeria working groups,
six-volume set / edited by Benjamin Ewa Ubi [and fourteen others].
Description: First edition. | Boca Raton : CRC Press, 2022. | Includes
bibliographical references and index. |
Contents: Volume 1. Fermentation and Algal Biotechnologies for the Food, Beverage and Other Bioproduct Industries — Volume 2. Agricultural biotechnology, biodiversity and bio-resources conservation and utilization — Volume 3. Medical Biotechnology, Biopharmaceutics, Forensic Science and Bioinformatics — Volume 4. Biosafety and Bioethics in Biotechnology — Volume 5. Microbiomes and Emerging Applications — Volume 6. Bioenergy and environmental biotechnology for sustainable development, | Summary: "This six volume book set examines a range of topics and applications related to biotechnology. Volumes include fermentation and algal biotechnologies; agricultural biotechnology; medical biotechnology, biopharmaceutics; biosafety, bioethics, biotechnology policy; microbiomes; bioenergy and environmental biotechnology for sustainable development"— Provided by publisher.
Identifiers: LCCN 2021047039 (print) | LCCN 2021047040 (ebook) |
ISBN 9780367766948 (v. 1 ; hardback) | ISBN 9781032013763 (v. 1 ; paperback) | ISBN 9780367766955 (v. 2 ; hardback) | ISBN 9781032014982 (v. 2 ; paperback) | ISBN 9780367766962 (v. 3 ; hardback) | ISBN 9781032015026 (v. 3 ; paperback) | ISBN 9780367767020 (v. 4 ; hardback) | ISBN 9781032015781 (v. 4 ; paperback) | ISBN 9780367767044 (v. 5 ; hardback) | ISBN 9781032018331 (v. 5 ; paperback) | ISBN 9780367767051 (v. 6 ; hardback) | ISBN 9781032018416 (v. 6 ; paperback) | ISBN 9781003178378 (v. 1 ; ebook) | ISBN 9781003178880 (v. 2 ; ebook) | ISBN 9781003178903 (v. 3 ; ebook) | ISBN 9781003179177 (v. 4 ; ebook) | ISBN 9781003180241 (v. 5 ; ebook) | ISBN 9781003180289 (v. 6 ; ebook)
Subjects: LCSH: Biotechnology—Nigeria.
Classification: LCC TP248.195.N6 M95 2022 (print) | LCC TP248.195.N6
(ebook) | DDC 660.609669—dc23
LC record available at https://lccn.loc.gov/2021047039
LC ebook record available at https://lccn.loc.gov/2021047040

ISBN: 978-0-367-76702-0 (hbk)
ISBN: 978-1-032-01578-1 (pbk)
ISBN: 978-1-003-17917-7 (ebk)

DOI: 10.1201/9781003179177

Typeset in Times
by codeMantra

Dedication

This book is dedicated to Professor Vincent Ado Tenebe, FBSN, Chairman of the Board of Trustees (BoT) of the Biotechnology Society of Nigeria (BSN) for his passion for Biotechnology and commitment to the cause of the Society.

Contents

Foreword

This book is an indispensable and timely project of the Biotechnology Society of Nigeria (BSN). It furnishes useful information on diverse areas of biotechnology and adjoining disciplines in a bid to equip academia, policies makers, farmers associations, non-government organizations and general public on safe application of biotechnology, biosafety and status of Nigeria's Biosafety Regulation, bioethics coupled with advocacy and human development. The book is written by the compendium of seasoned academics, researchers, high-ranking officers in relevant fields of government agencies and advocators with the cardinal goal to demystify the misconceptions that serve as a cog in the wheel of national development vis-à-vis application and adoption of modern biotechnology in achieving food security, safe healthcare delivery, partnership with industry for delivery of public goods to fast-track the emergence of a sustainable bioeconomy in Nigeria and safe environments that engender conservation and sustainable utilization of our biodiversity.

The first chapter of the book gives an insight of the status of Nigeria's Biosafety Regulation which includes efforts and regulatory instruments put in place by the Federal Government of Nigeria through the National Biosafety Management Agency (NBMA) in fulfilment of these international obligations to ensure that the application of modern biotechnology and the handling and use of genetically modified organisms are not harmful to the conservation and sustainable utilization of biodiversity.

Furthermore, reducing imprecision and unintended effects of classical recombinant DNA technology informed the emergence of genome-editing tools such as CRISPR/Cas9, TALENs and ZFNs. The second chapter of the book focuses on the most important genome-editing system – CRISPR/Cas9, provides its mechanism, summarizes regulations of genome-edited agricultural products in some countries and suggests a review of Nigerian process-triggered GM regulations to make its stringency dependent on the nature of the final genome-edited product. The importance of gene editing in agriculture and medicine has been demonstrated. However, global regulatory approaches to protect man and environment from potential risks associated with gene editing differ. The third chapter provides an overview on the stance of the Nigerian government and procedures in the regulation of gene-edited products. Also, it is shown that the use of gene-drive technology to produce transgenic mosquito is a sustainable tool to achieve malaria control from the vector angle. The fourth chapter discusses regulation of the emerging technology in Africa and emphasizes the roles of international organizations and continental agencies in ensuring safe applications of transgenic mosquitoes. Since biotechnology entails different techniques that include the usage of numerous biological organisms, processes and systems, chapter 5 discusses ethical and social aspects of modern biotechnology, including merits and demerits of biotechnological applications of humans and with four basic principles of bioethics.

The urgency for the need to adopt biotechnology in Nigeria was the main trigger for the adoption of the biotechnology policy in April 2001 and later in November of the same year, by the Federal Government of Nigeria which led to the full establishment and operation of the National Biotechnology Development Agency (NABDA), which I was opportuned to lead as the Director General/Chief Executive Officer for an active period of 8 years between 2005 and 2013. Chapter 6 seeks to elaborate on the different levels of public engagements with stakeholders, media and social media engagements held so far and their effectiveness in science communication. Moreover, meeting critical needs of development such as generation of products and services that meet human needs in food production, health services provision and maintenance of environmental integrity is the main goal of embracing modern biotechnology in Nigeria. The seventh chapter emphasizes more collaborative research and partnership with industry to ensure the delivery of public goods derived from the technology and to fast-track the emergence of a sustainable bioeconomy in Nigeria. Chapter 8 centees on communicating science to achieve the goal of fostering greater understanding of science and scientific innovations and gaining greater insight into diverse public views and concerns about the science related to a contentious issue involved in public controversy. It also presents researches built on evidence-based outputs useful for making decisions about how to communicate science most effectively to cover the knowledge gaps.

Chapter 9 provides detailed information on necessary regulatory risk evaluation frameworks for products that are developed using biotechnological approaches. It also throws light into necessary relationships that occur among regulatory policy, science and risk assessment. In Chapter 10 of this book, a chronicle of National Master Plan for biotechnology development in Nigeria was heighted. This lent credence to the approval of the National Biotechnology Policy in 2001 and the eventual establishment of NABDA. Almost all the universities in Nigeria offer biotechnology trainings at both undergraduate and postgraduate levels in Biological Sciences, Agriculture, Medical Sciences, Pharmaceutical and Health Sciences. The need for government to offer more incentives to drive biotechnology education in Nigeria, with more funding provided at all levels but especially at the postgraduate level, was also emphasized.

Chapter 11 explores the challenge of Science, Technology, Engineering and Mathematics (STEM) in technological development of Nigeria. The authors asserted that recognition of the challenges and the means of improving the current situation will go a long way in enhancing development in Nigeria. Causes of technological backwardness in Nigeria were also itemized. However, education of youths, provision and maintenance of infrastructural facilities, wealth creation from the national mineral resources, the limitation of existing items, and engagement in industrial espionage are the promising ways to ensure Nigeria's technological development.

The authors highlight both public and private organizations that have played prominent roles in capacity building in modern biotechnology in Chapter 12. Foundation for African Development through International Biotechnology (FADIB) was the first to mount yearly train-the-trainer workshops on DNA manipulating techniques in Nigeria; UNAAB (now FUNAAB), DANIFOL Biotechnology Consult Ltd, IITA Bioscience and Applied Biotech International Nigeria Ltd (ABINL) are prominent private sector players in these efforts. All these combinations have resulted in the change in the attitude of Nigerians to biotechnology and to the point that there are at least three GM crops approved for general use in Nigeria.

In the last chapter of this book, the authors point out that translating the abundant bioresources in Africa into value-added bio-products and services needs appropriate infrastructure and human resource development. They also highlight biotechnology education and skilled workforce trends for the bioeconomy, a curriculum framework, and proffer innovative approaches to rapidly build up competent biotech entrepreneurs and trainers in an emerging economy. Regional bioeconomy clearing houses and national commissions housed by the Presidency in African countries to facilitate multilateral cooperation and partnerships for rapid and sustainable growth of bio-based economies are recommended.

In a nutshell, I make bold to aver that recent pandemic caused by a novel Coronavirus – the SARS-COV-2 – upholds the veracity that biotechnology is one of the primary drivers of the global future. And virtually everything you need to know in biotechnology in nexus with adjoining disciplines is addressed in this book. Therefore, I recommend this book to be read by all.

Prof. Bamidele Ogbe Solomon, FBSN, Fmr. DG/CEO,
National Biotechnology Development Agency (NABDA)

Preface

In recognition of biotechnology's pivotal role as a burgeoning technology in fostering a knowledge-driven bio-based economy, the Biotechnology Society of Nigeria (BSN) has – in the about 40 years of its founding – made enormous contributions towards expanding the frontiers of biotechnology in Nigeria for sustainable livelihoods. This book is one of the multi-volume series on the Multidisciplinary Applications and Advances in Biotechnology: Contributions from the BSN, with each volume covering specific contemporary topics targeted at providing scientists in the academia, industry, research institutes/agencies, policy makers, etc. with recent advances in the field to bridge the knowledge gap in the crucial areas of biotechnology.

This book is designed to document advances in biotechnology in Nigeria over the past decades and the emerging biotech sector as a key driver of the bioeconomy. It is a timely knowledge product detailing key issues on recent developments in the field of biotechnology, particularly in Nigeria. It is written for use by various readers including postgraduate students, professionals in the area of research, industry players, policy makers, educationists, science advocacy groups, communication experts, etc. Additionally, the book engenders a deeper understanding of underlying developments in biotechnology, its historical perspectives and context in Nigeria, processes and science-policy nexus that have placed Nigeria as a leader in biotechnology in Africa.

This volume covers 3 of the 16 Working Groups of the Society, namely Biosafety, Bioethics and Legal Issues in Biotechnology (WG05); Biotechnology Policy Advocacy, Enlightenment and Public Engagement with Stakeholders/Policy Makers (WG06); and Biotechnology Education and Curriculum Development (WG15). It contains 12 chapters that bring to light the applications of available knowledge and deliverables in the thematic areas covered in the topics by the array of seasoned contributing authors. The chapters in the Biosafety and Bioethics section cover a wide range of topics including the status of Nigeria's biosafety regulation; regulatory considerations to ensure safety of gene-edited products in Nigeria such as those developed by CRISPR/Cas9 technologies; regulation of emerging technologies such as gene drive for transgenic mosquito control in Africa; ethical and social aspects of modern biotechnology; etc. In the Biotechnology Policy Advocacy, Enlightenment and Public Engagement with Stakeholders/Policy Makers section, the book covers topics such as Science Communication in Agricultural biotechnology and Public Engagement with Stakeholders in Nigeria; biotechnology development in Nigeria for meeting critical developmental needs; communicating scientific innovations based on a case study of agricultural biotechnology in Nigeria; and a review of biotechnology policies and regulatory frameworks for biotechnology advancements. Furthermore, in the Biotechnology Education and Curriculum Development section, the book covers topics such as the current status of biotechnology manpower development in Nigeria; challenges of Science, Technology, Engineering and Mathematics (STEM) in technological development of Nigeria; capacity building in contemporary biotechnology in Nigeria: history, impact and way forward; and an outlook on bioeconomy in Africa: strategies for transforming biotech knowledge and technical base into skilled workforce and productivity.

Each chapter addresses the relevant points in the subject areas covered and discusses each of the thematic areas concisely. The chapters are co-authored by a team of different scientific experts in the field which guarantees in-depth scholarly output in a high level of interdisciplinary collaborative manner. We have leveraged the pool of expertise available in the three Biotechnology Society of Nigeria (BSN) Working Groups to ensure a higher level of sustainability on the deliverables expected from these thematic research areas. We envisioned that this would foster a drive towards the rapid advancements in biotechnology in Nigeria; a connect between scientists and policy makers; biotechnology curriculum at the undergraduate and graduate levels; skilled manpower development and capacity building/training for a thriving biotech workforce; and deployment of the technology towards attaining the overarching goal of a sustainable knowledge-driven bio-based economy.

This book is a useful reference material, a comprehensive description of the recent advances in the field, and a valuable guide to the subject matter's original literature. It is specifically targeted at scientists and researchers working in the different areas of biotechnology including policy makers, educationists, communication experts, science advocacy groups, manpower development and training of biotech students in the universities and other tertiary institutions including the research institutes, biosafety regulation for genetically engineered organisms and other emerging technologies such as genome editing and gene drives, bioethics, etc. This book will be valuable to professionals in universities, research institutions, industry or government agencies, and policy makers/science advocacy groups.

Overall, this volume is an excellent piece of scientific contributions that targets a wide range of audiences including those involved in biotech manpower development/ training, global leaders working on fostering development in biotechnology and its global adoption in line with the biosafety regulatory framework in Nigeria, science communication experts, advocacy groups and policy makers. It should also be of relevance to industrialists with interest in developing Africa's bioeconomy and biotech start-ups, agricultural biotechnology sector, professionals involved in science communication and engagement with stakeholders, biosafety and bioethics professionals, farmers, policy makers, educators and researchers, and educational and research institutions. Readers will find this book a valuable resource containing useful information in the thematic areas contributed by a seasoned and diverse array of scholars within this professional Society.

The carefully selected editorial team and the authors of individual chapters drawn from several key institutions were chosen for their recognized international expertise and contributions to the various subject areas. Their willingness to make these invaluable contributions in their respective fields of expertise reflecting the different Working Groups and in a collaborative way to foster research networking is gratefully acknowledged. The editors wish to thank the different manuscript reviewers for their unceasing dedication and significant input. We thank Prof. Malachy O. Akoroda of the Department of Agronomy, University of Ibadan, who shared the initial idea that inspired this book initiative with me just after the conclusion of the 32nd Annual Conference of the Society held at the International Institute of Tropical Agriculture

(IITA) Ibadan, Nigeria. Furthermore, we highly appreciate the foresight, diligent support and understanding of the publisher, Taylor and Francis Group (USA), for publishing *Multidisciplinary Applications and Advances in Biotechnology – IV: Biosafety, Bioethics, Biotechnology Policy and Advocacy, and Capacity Building* with their international hallmark of excellence. We return special thanks to Dr. Marc Gutierrez and Mr. Nick Mould, whose tremendous and painstaking efforts, constant suggestions and guidance at all stages led to the successful publication.

Prof. Benjamin Ewa Ubi
President, Biotechnology Society of Nigeria (BSN)
& Chief Editor
26 February, 2021

Editors

Nwadiuto (Diuto) Esiobu (Ph.D., University of Louvain, Belgium; Post Doc. Massachusetts Institute of Technology, Cambridge, USA) is currently a Professor of Microbiology and Biotechnology at Florida Atlantic University, Boca Raton, FL, USA, where her microbial biotech laboratory integrates environmental genomics and health risk assessment with rapid pathogen detection using synthetic biology. Her research interests in agriculture and health include molecular ecology of plant invasion, soil fertility and the applications of microbiomes in humans and plants. She has served in senior professional leadership roles in academia, industry and policy, including selection panel for the US President's award for excellence in Math and Science (Washington, DC) and more than 11 years on Committees of the Board of the American Society for Microbiology. A Fellow of the Biotechnology Society of Nigeria and a Jefferson Science Fellow, she was selected by the National Academy of Sciences as a senior science advisor to Secretary Clinton's office of Global Food Security (GFS) and the Office of International Health and Biodefense (OES/IHB), US Department of State, Washington, DC, and named world-class Faculty by the City of Fort Lauderdale for her contributions to the scientific enterprise.

Dr. Arinze Stanley Okoli is an Associate Professor at Genoek – Centre for Biosafety, Universitetet II, Breivika, Tromsoe, Norway. His academic trainings include M.Sc. (Microbiology) – University of Nigeria, Nsukka, Nigeria; Ph.D. (Medical Sciences) – University of New South Wales, Sydney, Australia; and Post Doc (Section for Virology & Molecular Biology) – Genoek, Tromsoe, Norway. Dr. Okoli has executed several funded research projects. His current research interest is on biosafety of the use of modern genome-editing technologies, in particular CRISPR/Cas. A fellow of several national and international academic societies including Biotechnology Society of Nigeria and Australian and Norwegian Proteomic Societies, Dr. Okoli has contributed to over 30 medium- to high-impact peer-reviewed articles (ORCiD: https://orcid.org/0000-0003-0256-0856), technical reports, several conference proceedings and conference presentations in 17 different countries. He is a member of the scientific expert committee of the Norwegian Food Safety Authority and the European Medical Agency.

Sylvia Uzochukwu, Ph.D. is a Professor of Food Science and Biotechnology at the Department of Plant Science and Biotechnology, Federal University, Oye- Ekiti, Nigeria. She holds a B.Sc. in Botany from the University of Nigeria Nsukka, and M.Sc. and Ph.D. in Food Science and Technology, from the same University. She was the Director of the Biotechnology Centre of the Federal University of Agriculture, Abeokuta from 2002 - 2009, Head of Department of Biotechnology of the Federal University Oye-Ekiti from August to November 2013, and Dean of the Faculty of Science at the same University, 2013 to 2017. She has been involved in capacity building in biotechnology in Nigeria, and has organized up to 20 short training courses in Molecular Biology and Biosafety between 2000 and 2018, and has attended about eight (8) short training courses in Molecular Biology in Europe, Asia and America. She has won many International and National research grants for biotechnology research, and currently works on unintentional effects of genetic modification on GM foods approved for use in Nigeria. She is currently the Director of the Biotechnology Centre, Federal University Oye-Ekiti, Nigeria.

Emeka Godfrey Nwoba, Ph.D., is currently a research Fellow at the Algae Research & Development Centre, Murdoch University, Western Australia, Australia. His research areas include industrial & environmental biotechnology, microalgae biotechnology, applied phycology/algology, energy-efficient photobioreactors (PBRs) and biostatistics. He has over 6 years of research and development experience in algae (microalgae and macroalgae) biotechnology including algae-based wastewater treatment, bio-based products from algae, and design, optimization and scale up of PBRs. His academic training was at Ebonyi State University, Abakaliki, Nigeria, where he received his B.Sc. and M.Sc. degrees, in Biochemistry/Industrial Biotechnology. Dr. Nwoba obtained his PhD in Biotechnology/Energy from Murdoch University, Australia. He has published several peer-reviewed research and review articles in Q1 Journals and book chapters. His current research focuses on partnering engineering, material science and biology to develop high efficiency/next-generation algal PBRs (i.e. autonomous/standalone PBRs) with the capability to co-produce chemical (algal biomass) and electrical energy towards freshwater-based cooling and grid-electricity independent operation. He is a member of Australasian Society of Phycology and Aquatic Botany, International Society for Applied Phycology, Phycological Society of Nigeria and Biotechnology Society of Nigeria.

Dr. Christpeace Nwagbo Ezebuiro started his early career in 2005 at the National Biotechnology Development Agency (NABDA) in Abuja and rose to the role of Head, Clean Technology Division with responsibilities that include development of technologies for water and wastewater management, contaminated-soil remediation, biomass energy production and development of strategies for advancing agricultural research products to market and industry. Dr. Ezebuiro is an expert in technology development and deployment, policy analysis and entrepreneurship.

Charles Oluwaseun Adetunji, Ph.D., is an Associate Professor of Microbiology and Biotechnology at the Microbiology Department, Faculty of Science, Edo State University Uzairue EDSU, Edo State, Nigeria, where he utilized the application of biological techniques and microbial bioprocesses for the actualization of sustainable development goals and Agrarian revolution, through quality teaching, research and community service. He is currently the Acting Director of Intellectual Property and Technology Transfer, Sub-Dean for Faculty of Science, and the Head of department of Microbiology at EDSU. He is a Visiting Professor and the Executive Director for the Center of Biotechnology, Precious Cornerstone University, Ibadan.. He has won several scientific awards and grants from renowned academic bodies like Council of Scientific and Industrial Research (CSIR), India; Department of Biotechnology (DBT), India; the World Academy of Science (TWAS), Italy;

Netherlands Fellowship Programme (NPF), the Netherlands; the Agency for International Development Cooperation; Israel; and Royal Academy of Engineering, UK, among many others. He has filed several scientific patents on Bioherbicides, Biopesticides, Nanobiosurfactants, Nanobiopesticdes and many more. He has published over 300 scientific articles in refereed journals and conference proceedings both international and local journals. He was ranked among the top 500 prolific authors in Nigeria between 2019 till date by SciVal/SCOPUS. He is currently editing several biotechnology text books with Elsevier. He was recently appointed as the President and Chairman Governing Council of the Nigerian Bioinformatics and Genomics Network Society. He is presently a series editor with Taylor and Francis, USA editing several textbooks on Agricultural Biotechnology, Nanotechnology, Pharmafoods, and Environmental Sciences. The breadth of his scholarly contributions to research is evident from his contributions which cover several topics relating to Food Security, Agriculture and Environmental Sustainability. His research interest includes microbiology, biotechnology, bioresources and nanotechnology. He is an editorial board member of many international journals and serves as a reviewer to many double-blind peer-review journals like Elsevier, Springer, Taylor and Francis, Wiley, PLOS One, Frontiers, Nature, American Chemistry Society, Bentham Science Publishers, etc. He is a member of many scientific and professional bodies including the American Society for Microbiology, Nigerian Young Academy, Biotechnology Society of Nigeria and Nigerian Society for Microbiology. He has won a lot of international recognition and also acted as a keynote speaker delivering invited talks/position papers at various universities, research institutes and several centres of excellence which span across several continents of the globe. Over the last 15 years, he has built strong working collaborations with reputable research groups in numerous and leading universities across the globe. He is the convener for Recent Advances in Biotechnology, which is an annual international conference where renowned Microbiologists and Biotechnologists come together to share their latest discoveries (https://raibconference.wixsite.com/website). He is the Founder & CEO of BECTIK Biotechnology and Nanotechnology Company. His Biotechnology Company consists of a team of leading academics in their fields and thrives to deliver solutions all around Bio- and Nanotechnology. The company also delivers solutions that are tailored to your business needs and for any scale.

Abdulrazak Ibrahim, Ph.D., is an agricultural biotechnologist with experience and skills on plant biotechnology stewardship, stakeholder engagement, foresight analysis, development of transgenic crops, laboratory and field management of GM crops, biosafety analysis, adoption and discontinuation of events. He was trained in the Brazilian Agricultural Research Cooperation (EMBRAPA Center for Genetic Resources), from where he obtained a doctorate degree in Molecular Biology from Universidade de Brasilia (UnB) and M.Sc. from Universidade Federal do Ceara (UFC), Fortaleza, and acquired best practices in agricultural biotechnology leading to the development and patenting of a transgenic technology for the control of whitefly using RNA interference technology within EMBRAPA, and the development of virus-resistant cowpea, under the Africa-Brazil Innovation Platform, implemented by EMBRAPA and Forum for Agricultural Research in Africa (FARA), with support from the Department for International Development (DFID), Bill & Melinda Gates Foundation (BMGF), International Fund For Agricultural Development (IFAD) and Foundation Arthur Bernardes (FUNARBE).

Benjamin Ewa Ubi, Ph.D., is a Professor of Plant Breeding and Biotechnology. He obtained his B.Sc. Honours (Agric.) Agronomy from the University of Calabar (1990) and M.Sc. Agronomy – Crop Science (Plant Breeding and Tissue Culture) (1994) and Ph.D. Agronomy (Plant Breeding and Biotechnology) (1998) from the University of Ibadan, Nigeria. His Ph.D. research was completed at the then Biotechnology Laboratory of the International Institute of Tropical Agriculture (IITA), Ibadan, where he was a graduate research fellow and became the President of the International Association of Research Scholars and Fellows (IARSAF) in 1996. Among several other research engagements, Prof. Ubi served as Science and Technology Agency (STA) Postdoctoral Fellow under the Japan International Science and Technology Exchange Centre (JISTEC) 1999–2001 at the Plant Breeding Department (Biotechnology Laboratory) of the National Institute for Livestock and Grassland Science (NARO, National Agricultural Research Organization), Nishinasuno, Japan; and as a three-term Visiting Scientist at the NARO National Institute of Fruit Tree Science (NIFTS), Tsukuba, Japan, under the auspices of the Inoue Research Foundation of Japan Fellowship (April 2003–April 2004) and the Japanese Society for the Promotion of Science (JSPS) visiting fellowships (October 2008–October 2009; May 2012–March 2013). Prof. Ubi served as Visiting Professor at the Arid Land Research Centre (The Global Centre of Excellence for Dryland Science) (Molecular Breeding Laboratory), Tottori University, Japan; and Visiting Professor University of Tokyo, Japan. Among several other academic engagements, Prof. Ubi has served as the Director of the Biotechnology Research and Development Centre, Ebonyi State University, Abakaliki (August 2006–March 2008) and the Head of the Department of Biotechnology, Ebonyi State University for several years; Dean, Faculty of Science, Ebonyi State University Abakaliki and Pioneer Editor-in-Chief of *Science Journal*, https://sciencejournal.ng, a publication of the Faculty of Science, Ebonyi State University, Abakaliki; President, Biotechnology Society of Nigeria (BSN) and Leader of the Crops Genetics, Breeding and Genetics Working Group of the Biotechnology Society of Nigeria (BSN).

Contributors

Charles Oluwaseun Adetunji
Applied Microbiology, Biotechnology and Nanotechnology Laboratory, Department of Microbiology
Edo State University Uzairue
Edo State, Nigeria

A. Akpa
National Biotechnology Development Agency (NABDA)
Abuja, Nigeria

Modupe Stella Ayilara
Food Security and Safety Niche Area, Faculty of Natural and Agricultural Science
North-West University
Mmabatho, South Africa

Scholastica Bello
National Biosafety Management Agency (NBMA)
Federal Ministry of Environment
Abuja, Federal Capital Territory

Ebiamadon Andi Brisibe
Department of Genetics & Biotechnology
University of Calabar
Calabar, Nigeria

Mwajim Bukar
North East Zonal Biotechnology Centre of Excellence
University of Maiduguri
Maiduguri, Nigeria

Joan Campbell
Department of Clinical Biochemistry
Frederiksberg Hospital
Frederiksberg, Denmark

Aliyu Daja
Biochemistry Department, Faculty of Science
University of Maiduguri
Maiduguri, Nigeria

Rufus Ebegba
National Biosafety Management Agency (NBMA)
Federal Ministry of Environment
Abuja, Federal Capital Territory

Nwadiuto (Diuto) Esiobu
Applied Biotech International Nigeria Ltd
Wuye, Abuja
and
Biological Sciences Department
Florida Atlantic University
Boca Raton, USA

N. C. Ezebuiro
National Biotechnology Development Agency (NABDA)
Abuja, Nigeria

Abubakar Gidado
North East Zonal Biotechnology Centre of Excellence
University of Maiduguri
Maiduguri, Nigeria

R. S. M. Gidado
Department of Agricultural Biotechnology
National Biotechnology Development Agency (NABDA)
Wuse, Nigeria

Abdulrazak Ibrahim
Department of Biochemistry
Ahmadu Bello University
Zaria, Nigeria

J. C. Igwe
Kaduna State University
Kaduna, Nigeria

Ivan Ingelbrecht
FAO/IAEA Laboratory on Plant Breeding and Genetics
Vienna, Austria

A. Isah
Department of Agricultural Biotechnology
National Biotechnology Development Agency (NABDA)
Wuse, Nigeria

S. O. O. Iweajunwa
Department of Agricultural Biotechnology
National Biotechnology Development Agency (NABDA)
Wuse, Nigeria

Paul Keese
Biotechnology Regulatory Services
Office of Gene Technology Regulator
Canberra, Australia

Olugbenga Samuel Michael
Cardiometabolic Research Unit
Department of Physiology
College of Health Sciences
Bowen University
Iwo, Nigeria

N. Muchiri
Department of Communication and Partnerships
African Agricultural Technology Foundation
Nairobi, Kenya

Alex Ochem
Cape Peninsula University of Technology
Cape Town, South Africa

Olu Odeyemi
Obafemi Awolowo University
Ile-Ife, Nigeria

James C. Ogbonna
Department of Microbiology
University of Nigeria
Nsukka, Nigeria

Arinze S. Okoli
GenØk Centre for Biosafety
University of Tromso
Norway

Olugbemi Tope Olaniyan
Edo State University Uzairue
Edo State, Nigeria

Julius Kola Oloke
Department of Pure and Applied Biology
Ladoke Akintola University of Technology
Ogbomoso, Nigeria

Joy Onwude
National Biosafety Management Agency (NBMA)
Federal Ministry of Environment
Abuja, Federal Capital Territory

Christie Onyia
Biological Sciences Department
Godfrey Okoye University
Enugu, Nigeria

Benjamin Ewa Ubi
Department of Biotechnology
Ebonyi State University
Abakaliki, Nigeria
and
Arid Land Research Center
Tottori University
Tottori, Japan

Kingsley Eghonghon Ukhurebor
Climatic/Environmental/
Telecommunication Physics Unit,
Department of Physics
Edo State University Uzairue
Edo State, Nigeria

Sylvia Uzochukwu
Biotechnology Centre
Federal University
Oye-Ekiti, Nigeria

Ivar Virgin
Stockholm Environment Institute
Stockholm, Sweden

1 Status of Nigeria's Biosafety Regulation

Rufus Ebegba, Scholastica Bello, and Joy Onwude
National Biosafety Management Agency (NBMA),
Federal Ministry of Environment

CONTENTS

DOI: 10.1201/9781003179177-1

1.1 INTRODUCTION

Nigeria as a country is endowed with a variety of plant and animal species forming Nigeria's biodiversity which is in abundance within the country, highly cherished and therefore, needs conservation and sustainable utilization (First National Biodiversity Report, 2001). At the 1992 Earth Summit in Rio de Janeiro, Brazil, world leaders agreed on a comprehensive strategy for "sustainable development" – meeting our needs for health, environment and biodiversity while ensuring that we leave a healthy and viable world for future generations. One of the key agreements adopted at Rio was Convention on Biological Diversity (CBD), which establishes three main goals: the conservation of biological diversity, the sustainable use of its components, and the fair and equitable sharing of the benefits arising from the use of genetic resources (CBD, 1992). Nigeria signed the CBD in 1992 to show commitment to Global Biodiversity conservation (UN, Biosafety Clearing House BCH).

Irrespective of the potential benefits of modern biotechnology, there are concerns that these products have potential adverse effects on human, plant and animal health, biological diversity and the environment. Also, there are several important socio-economic, cultural and ethical issues to be considered in the adoption and use of the products of modern biotechnology. As a result of these concerns, the Cartagena Protocol on Biosafety (CPB) was adopted in Montreal on 29 January 2000 at an extraordinary meeting of the Conference of the Parties to the CBD. The CPB makes provisions to regulate, manage or control risks associated with transfer, handling and use of organisms and derived products resulting from modern biotechnology that may have adverse effects on conservation and sustainable use of biological diversity, taking into account risks to human health, focusing on their transboundary movement (CPB, 200).

Nigeria signed the CPB in 2000 and ratified it in 2003 and subsequently, the Nagoya Kuala Lumpur Supplementary protocol on Liability and Redress to the Cartagena Protocol in Biosafety in 2010 (UN Biosafety Clearing House, BCH). As a signatory to the CBD and CPB, Nigeria is obliged to implement the articles of the CPB and develop its own national regulatory framework for the safe transfer, handling, use and release of genetically modified organisms (GMOs) and products resulting from modern biotechnology. In fulfilment of these international obligations, the Nigerian government signed into Law the National Biosafety Management Agency Act, 2015 on the 18th of April 2015. This Act established the National Biosafety Management Agency (NBMA), charged with the responsibility of providing regulatory framework, institutional and administrative mechanism for safety measures in the application of modern biotechnology and the use and handling of GMOs in Nigeria. This is with a view to preventing any adverse effect on human health, animals, plants and the environment.

The Government of Nigeria, in keeping with the dynamics of global modern technological developments, endorsed the NBMA (amended) Act 2019 in August 2019, which expanded the mandate of NBMA to include emerging biotechnology applications (gene drive, synthetic biology, gene editing) as well as biosecurity.

1.1.1 Mandate, Vision and Mission

The mandate, vision, mission, core values and enabling authority of the NBMA are as follows:

Mandate: Provision of regulatory framework, institutional and administrative mechanisms for safety measures in the application of modern biotechnology in Nigeria with a view to preventing any adverse effects on human health, animals and the environment.

Vision: Having in place a holistic, responsive and vibrant National Biosafety System that offers the best of its potentials in the regulation of modern biotechnology activities in Nigeria, with a view to preventing any adverse impact on the conservation and sustainability of biodiversity taking into account risk to human health.

Mission: To promote the basic tenets of biosafety as enunciated in the Cartagena Protocol on Biosafety and enforce Nigeria National Biosafety Management Agency Act 2015, regulations and guidelines to ensure safe application and use of products of modern biotechnology.

Core values: The Agency, as the Competent National Authority on Biosafety issues, operates on the following core values: safety to the environment and human health, respect for human dignity, transparency, integrity and professionalism.

Enabling authority: Established by the National Biosafety Management Agency Act 2015, to provide regulatory framework to adequately safeguard human health and the environment from potential adverse effects, while harnessing the potentials of modern biotechnology and its derivatives, for the benefit of Nigeria.

The Agency is governed by a Board. The day-to-day activities of the Agency are under the direction of the Director General/Chief Executive Officer. The Agency has a staff strength of 353.

1.1.2 Problem Statement and Justification

Modern biotechnology is believed to possess the potentials to help improve food security, promote sustainable use of natural resources and enhance agricultural productivity by introducing resistance to pests and disease without costly purchased inputs; heighten crops' tolerance to adverse weather and soil conditions; improve the nutritional value of some foods; and enhance the durability of products during harvesting or shipping. However, the use of the techniques of modern biotechnology in agriculture has given rise to ardent debates over the last 30 years about the benefits versus the risks posed by GMOs and their products thereof (James, 2011). Toxicity, allergenicity, pathogenicity, gene flow and effects on non-target organisms are just a few concerns regarding GMOs. It is to this end that biosafety has come to address these concerns. There is need for effective communication of Biosafety Regulation in Nigeria so as to allay the fears of her citizens and equip them with the knowledge of the role of the government in ensuring adequate protection of human health, biodiversity and the environment.

This chapter intends to give an overview of the status of Biosafety Regulation in Nigeria. In this chapter, particular attention will be given to the following objectives:

- To state the various biosafety regulatory instruments developed;
- To give an outline of the biosafety decisions reached;
- To outline the challenges faced by the Agency in the discharge of her duties.

1.2 EVOLUTION OF BIOSAFETY REGULATION IN NIGERIA

Nigeria has evolved through various biosafety regulation stages as enumerated in Table 1.1 with dates.

1.3 ESSENTIALS OF THE NIGERIAN BIOSAFETY SYSTEM

1.3.1 Risk Analysis

The NBMA uses the National Biosafety Risk Analysis Framework, 2018, to provide guidance on the rationale and approach to risk analysis and to enable a consistent and rigorous risk analysis approach to evaluating Permit applications of GMOs.

Risk Analysis integrates the assessment, management and communication of risks posed by, or as a result of, modern biotechnology (OGTR, 2013). All applications for Permit dealings with GMOs require case-by-case assessment and preparation of a risk assessment and risk management plan by the applicant. Details of the GMO and the proposed activities, including any proposed controls, limits or containment measures form the specific context for the risk assessment and risk management plan. Details of the parent organism and the environment where activities with the GMO will occur form the comparative baselines for the risk assessment.

The Risk Context defines the parameters within which risk is assessed, managed and communicated, defines the scope and boundaries, and sets the broad terms of reference and criteria against which the significance of risk will be evaluated, as well as the structure and processes for the analysis. The purpose of risk assessment is to identify and characterize risks to the health and safety of people or to the environment from dealings with GMOs, posed by or as the result of modern biotechnology (National Biosafety Risk Analysis Framework, 2018).

Risk assessment identifies risks from plausible sets of circumstances that may result in harm to people or to the environment and estimating the level of risk on the

TABLE 1.1
Evolution of Biosafety Regulation in Nigeria

Regulation Stages	Dates
First National Biosafety Guidelines, under Federal Ministry of Agriculture and Natural Resources	1994
Biodiversity/Biosafety activities under the defunct Federal Environmental Protection Agency (FEPA)	1997
National Biosafety Unit, Federal Ministry of Environment	1999
Second National Biosafety Guidelines, under Federal Ministry of Environment (FMEnv)	2001
Nigeria signed and ratified the Cartagena Protocol on Biosafety	2000, 2003
Development of National Biosafety Framework (NBF)	2002–2006
Development of National Biosafety Policy	2002–2006
Draft Biosafety Bill, among others	2002–2006
Nigeria signed the Nagoya Kuala Lumpur Supplementary protocol to the Cartagena Protocol on Biosafety (not yet ratified)	2012
Development of other Biosafety Guidelines, and Draft Biosafety regulations	2012–2015
National Biosafety Management Agency Act 2015	2015
Establishment of the National Biosafety Management Agency (NBMA) 2015	2015
Review of draft Biosafety Regulations and Guidelines	2015–2017
Operationalization of NBMA Act	2015–date
Approval of National Biosafety Policy	2017
Approval of Biosafety Regulations	2017
Adoption of Biosafety Guidelines	2018
Appointment of Board Members	2018
Amendment of NBMA Act 2019 leading to the expansion of the mandate of NBMA to include regulation of synthetic biology, gene drive, gene editing and biosecurity	2019
Drafting (gene editing, gene drive, biosecurity policy and emerging response strategy) of other biosafety guidelines in line with expanded mandate	2019–2020

basis of the seriousness and likelihood of harm. Risks are identified by considering a broad range of circumstances whereby the proposed dealings with a GMO are postulated to have the potential to cause harm to people or to the environment through a plausible causal pathway between the GMO and an adverse outcome. Risks are then characterized in terms of the degree of seriousness and likelihood of potential harm, which are combined to estimate the level of risk as negligible, low, moderate or high. The risk assessment initially considers a wide range of possible risks, but puts great emphasis on more substantive risks, which receive more detailed characterization. Risks that are estimated to be greater than negligible are then considered by risk management for control or mitigation (National Biosafety Risk Analysis Framework, 2018).

Risk management evaluates, selects and implements plans or actions to ensure that risks are appropriately managed. The purpose of risk management is to protect the health and safety of people and to protect the environment by controlling or mitigating risk. Risk management involves prudent judgements about which risks require management (risk evaluation), the choice and application of treatment measures, and ultimately, whether the dealings with GMOs should be permitted. Risk management includes preparation of a risk management plan by evaluating and treating risk, applying general risk management measures and proposing Permit conditions to give effect to management measures. In addition, risk management includes monitoring and reviewing to provide feedback on all steps in risk analysis and ensure the outcomes remain valid for future findings or changes in circumstances. Risk management evaluates risks that may warrant control measures and determines the appropriate Permit conditions to manage risk (National Biosafety Risk Analysis Framework, 2018).

The risk assessment and risk management plan forms the basis upon which NBMA decides whether to issue a Permit. To issue a Permit, the NBMA must be satisfied that risks can be managed to protect human health and safety and the environment. If the NBMA considers that risks posed by proposed dealings with a GMO cannot be managed, a Permit would be refused (National Biosafety Risk Analysis Framework, 2018).

Risk communication is integral to the processes of risk assessment and risk management and involves development of an interactive dialogue between the NBMA and stakeholders to provide open, transparent and consultative risk-based regulation of GMOs. The National Biosafety Committee (NBC) undertakes extensive consultation with a diverse range of expert groups, authorities and key stakeholders, including the public, before deciding whether to recommend for Permit. In many instances, differing perceptions of risk can influence the approach of stakeholders to particular issues. The NBMA can also seek advice on ethical and social issues raised by modern biotechnology from the NBC. The NBMA endeavours to provide accessible information to interested parties on applications, Permits, dealings with GMOs, trial sites and the processes of risk assessment, risk management, monitoring and compliance undertaken by the NBMA (National Biosafety Risk Analysis Framework, 2018).

The Risk Analysis Framework is part of the NBMA commitment to clarity, transparency and accountability of decision-making processes (National Biosafety Risk Analysis Framework, 2018).

1.3.2 Nigeria's Biosafety Regulatory Instruments

These instruments are products of deliberations by a range of Government Ministries, Departments and Agencies (MDAs) and non-governmental agencies, as well as consultations with other relevant stakeholders. These Instruments include, the NBMA Act as amended, National Biosafety Policy, Regulations and Guidelines (NBMA, 2017).

1.3.2.1 National Biosafety Management Act 2015

The National Biosafety Management Act 2015 prescribes procedures for the application of the use of modern biotechnology before the adoption and use of any GMOs as well as penalties for contravening the Biosafety Act. The Biosafety Act is therefore a safety valve for harnessing the potentials of modern biotechnology.

The Act seeks to provide derived benefits from modern biotechnology under a legal framework for economic growth, improved agriculture, job and wealth creation, industry growth and sustainable environment; minimize risks to human health; confirm and harness the potentials of modern biotechnology; protect and guard against any adverse effect of GMOs on biological diversity and the environment; guard against any socio-economic consequences; give confidence in the practice of modern biotechnology, use and handling of GMOs and GM products; reaffirm Nigeria's commitment to the principles of international agreements and treaties (CBD and in particular the CPB); determine in advance when hazards to human health and natural systems will result if any particular GMO is released into the environment amongst others; and carry out proper regulation of imported GM products.

1.3.2.2 National Biosafety Regulation 2018

The objectives of these Regulations are to complement and enhance the provisions of the Act; provide details of regulatory and supervisory requirements necessary to promote and aid the efficient and profitable implementation of the provisions of the Act; and facilitate the attainment of the goals for which the Agency is established in Nigeria.

Some of these biosafety regulations are: GMOs' Import, Export and Transit; GMOs' Packaging, Identification and Transport; GMOs' Commercial Release; Biosafety Liability and Redress; and GMOs' Contained Use and Confined Field Trial (CFT).

1.3.2.3 National Biosafety Policy 2018

The policy sets out objectives, strategies and implementation procedures for a range of government-led activities, which together create the framework for a national biosafety regime. It addresses the safe use, transportation, storage and handling of GMOs – including requirements for transboundary movement – and sets a policy framework for supporting research and public education on modern biotechnology. The policy is designed to meet not only international obligations, specifically those set out in the CPB, but also the peculiar needs and requirements of Nigeria as it seeks to benefit from the advantages of the technology.

1.3.2.4 National Biosafety Risk Analysis Framework 2018

The Risk Analysis Framework provides guidance on how the NBMA carries out risk analysis of GMOs in accordance with the Act.

1.3.2.5 Biosafety Guidelines

1.3.2.5.1 National Guidelines for Institutional Biosafety Committee 2018

These guidelines set out the procedures to be followed by any institution (research institutes, universities and polytechnics, international research centres, industrial research and development units which plan to undertake modern biotechnology research and/or development) in the constitution of the Institutional Biosafety Committee. This committee in cooperation with the NBMA acts as a statutory oversight for all modern biotechnological research being carried out in the institution.

1.3.2.5.2 Biosafety Containment Facility Guidelines 2018

The guidelines contain the requirements for certification of Biosafety Physical Containment Level 2 (PC2) Facility.

1.3.2.6 Biosafety Manuals

1.3.2.6.1 Biosafety Laboratory Manual 2018

Biosafety Laboratory Manual guides scientists in the correct laboratory work ethics. Biosafety Laboratory Manual serves as the national code of practice for the safe handling of GMOs in laboratories. The manual has been developed for handling organisms which are unlikely to pose serious hazards to laboratory workers, the community, livestock or the environment.

1.3.2.6.2 National Biosafety Administrative Manual, 2018

This manual serves as a step-by-step guide for biosafety personnel with regard to handling applications for various biosafety services. It specifies the processes involved from the receipt of an application right through to when decision(s) is taken and communicated to the applicant. The manual serves as a working tool for personnel of the NBMA as well as a guide for prospective applicants.

Other documents include: National Biosecurity Policy and Action Plan, National Guidelines on Gene Editing, Guideline for the Importation of Genetically Modified Organisms for Food Feed and/or Processing, National Biosafety Application Processing Manual, Decision Document, National Biosafety Management Information Guide, National Biosafety Communication Strategy, National Biosafety Emergency Response Strategy, Cessation Order, Revocation Order, Nigeria National Guidelines on Biosafety Socio-economic Considerations in Decision-Making Process, Confined Field Trial Monitoring and Inspection Manual, Biosafety Application Form, GMOs Import/Shipment Form, Accreditation of Institute Application Form, and Certification of Biosafety Containment Facility Form. The Agency is currently drafting National Guidelines for Gene Drive and National Guidelines on Genetically Modified Fish.

1.3.3 National Genetic Modification, Detection and Analysis Laboratory

The NBMA possesses a well-equipped genetic modification detection and analysis laboratory (*a view inside the laboratory is shown in Figure 1.1*) to checkmate any risks associated with the technology which is headed by an expert in microbiology and plant biotechnology.

1.4 GOALS OF THE NIGERIA BIOSAFETY SYSTEM

The goals of the Nigerian Biosafety System are outlined as follows: maximize benefits and minimize risks; ensure safety to humans and animals; ensure safety to the environment and agriculture; independent check on products; carry out effective risk assessment and risk management; make decisions based on the best available science; build public confidence in new and emerging technologies; ensure biosecurity; and alleviate public concerns.

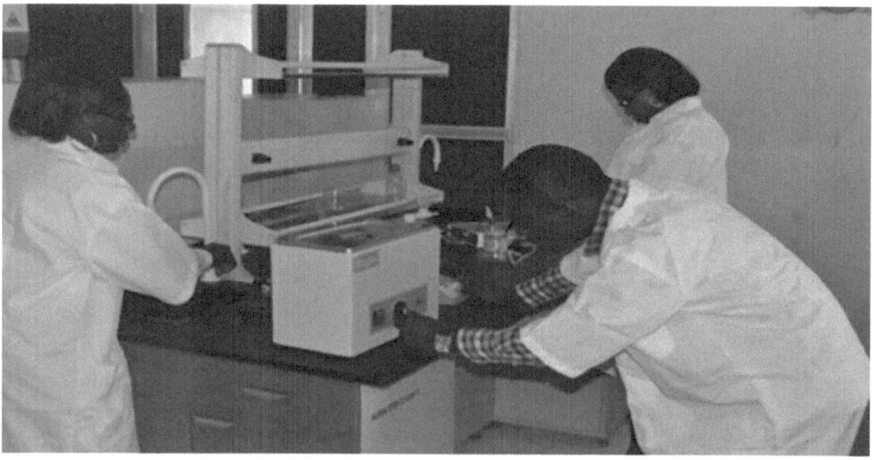

FIGURE 1.1 Showing experts carrying out an analysis in the NBMA GM detection and analysis laboratory.

1.5 NIGERIA'S BIOSAFETY DECISIONS SO FAR

1.5.1 Accreditation of Institutes to Carry Out Modern Biotechnology Activities

The following institutes have been accredited to carry out activities involving modern biotechnology:

- Federal University of Technology (FUTA), Akure;
- Institute for Agricultural Research (IAR), Zaria;
- National Biotechnology Development Agency (NABDA), Abuja;
- National Cereals Research Institute (NCRI), Badeggi;
- National Root Crops Research Institute (NRCRI), Umudike;
- International Institute of Tropical Agriculture (IITA), Ibadan;
- University of Calabar, Calabar.

1.5.2 Approval for Confined Field Trials in Nigeria

Some field trials in Nigeria have been approved for planting of GMOs in confinement, some have been concluded while some are on-going, and they include:

- Biofortified cassava enhanced with pro-vitamin A at National Root Crops Research Institute, Umudike;
- Biofortified cassava enhanced with iron at National Root Crops Research Institute, Umudike;
- Genetically modified Cassava engineered to express elevated levels of iron and zinc in the storage roots at National Root Crops Research, Umudike;
- Cowpea modified for resistance to Maruca insect pest at Institute of Agricultural Research (IAR), Zaria (Figure 1.2);

FIGURE 1.2 Showing GM cowpea under confined field trial.

FIGURE 1.3 Showing GM sorghum under confined field trial.

- Africa Biofortified Sorghum: bioavailability of iron, zinc, protein and pro-vitamin A at Institute of Agricultural Research (IAR), Zaria (Figure 1.3);
- Genetically modified rice modified for Nitrogen use efficiency, water use efficiency and salt tolerance at National Cereals Research Institute, Badeggi;
- Genetically modified Cassava resistant to cassava mosaic virus and brown streak virus at National Root Crops Research Institute, Umudike (Figure 1.4);
- Maize resistant to insect and herbicide tolerance for Monsanto Agriculture Nigeria Ltd (yet to commence);
- Genetically modified cassava for post-harvest starch deterioration reduction at IITA, Ibadan;

FIGURE 1.4 Showing GM cassava under confined field trial.

- Herbicide tolerance soybean at National Biotechnology Development Agency, Abuja;
- Genetically modified cassava for increase in starch content at IITA, Ibadan;
- Water efficient Maize for Africa (WEMA) for drought tolerant and stem borer insect pest resistance at Institute of Agricultural Research (IAR), Zaria.

1.5.3 Approval for Commercial Release

The approved crops for commercial release are:

- Bt. Cotton for Monsanto Agriculture Nigeria Ltd;
- Bt. Cowpea for IAR, Zaria.

1.5.4 Approval for Food, Feed and/or Processing (FFP)

Some GM grains were approved to be imported for food, feed and/or processing and also vegetable oil, and they are:

- Permit of GM maize for feed processing to Flour Mills Nigeria Plc. (Golden Agri Inputs Ltd);
- Importation of nine (9) events of GM Maize: lines Bt11, GA21, MON810, NK603, DAS1507, MIR162, MIR604, MON89034, and MON88017 for feed processing to WACOT Nig. Ltd;
- Importation of GM soybean for feed processing to Agboola Farms Ltd;
- Importation of GM soybean for feed processing to CHI Farms Ltd;
- Importation of GM Soybean for feed processing to Elephant Group Ltd;
- Importation of GM Soybean for production of edible oil to ROM Oil Mills Ltd;

- Importation of GM Maize for feed processing to Flour Mills Nigeria Plc. (Golden Agri Inputs Ltd);
- Importation and use of the following GM Soybean lines: 356043, A5547-127, MON87701, MON87708 and MON89788 for feed and edible oil processing to Grand Cereals Ltd;
- Importation and use of GM Maize lines: TC1507, MON88017, GA21 and MIR604 for poultry feed processing to Valency Cashew Processing Ltd;
- Use of GM animal pet foods containing GM Maize to Game Discount World Nigeria Ltd.

1.5.5 Gene Registration

Some approvals have been granted to the following technology developers for gene registration:

- Syngenta as the Developer of Genetically Modified Maize (3272 and MZIR093);
- Syngenta as the Developer of Genetically Modified Soybean (SYHTOH2);
- Syngenta, South Africa (Pty) Limited as the Developer of Genetically Modified Maize Bt.11/PAT;
- Monsanto Agriculture Nigeria Limited as the Developer of Genetically Modified Soybean Events: MON89788-1, MON87769-7, MON87708, MON87705-6, MON87701-2, and MON04032-6/PAT;
- Monsanto Agriculture Nigeria Limited as the Developer of Genetically Modified Maize Events: MON87460-4, MON874-27-7, MON 00810-6, MON00603-6, MON89034-3, and MON-88017-3/PAT.

1.6 SYNERGY AMONG STAKEHOLDERS FOR EFFECTIVE REGULATION OF MODERN BIOTECHNOLOGY AND GMOs IN NIGERIA

Recognizing the importance of synergy among stakeholders, the NBMA is collaborating with several MDAs, research institutes, NGOs, academia, foreign/developmental partners, Media, etc. in carrying out its mandate of ensuring that the practice of modern biotechnology and use of GMOs are safe to human health and to the environment.

1.6.1 Collaborating Organizations in Nigeria with NBMA

1.6.1.1 Foreign/Developmental Partners

- Africa Agricultural Technology Foundation (AATF), Kenya/Nigeria;
- African Biosafety Network of Expertise (AU-NEPAD ABNE), Dakar;
- Argentine Embassy/Argentine National Advisory Commission of Agricultural Technology (ONABIA);
- Danforth Plant Science Center, USA;

- Food and Agricultural Organization (FAO);
- International Centre for Genetic Engineering and Biotechnology (ICGEB), Italy;
- International Food Policy Research Institute (IFPRI), USA;
- International Institute of Tropical Agriculture (IITA), Ibadan;
- International Service for the Acquisition of Agri-Biotech Applications (ISAAA), Kenya;
- Michigan State University (MSU);
- Missouri State University (USA);
- Office of the Gene Technology Regulator (OGTR), Australia;
- Open Forum on Agricultural Biotechnology (OFAB), Kenya/Nigeria;
- Program for Biosafety Systems (PBS), USA;
- UNEP/GEF, Kenya (UN);
- United Nations Industrial Development Organization (UNIDO);
- United States Agency for International Department (USAID);
- United States Department for Agriculture (USDA);
- World Health Organization (WHO).

1.6.1.2 Ministries, Department and Agencies

- Federal Ministry of Agriculture & Rural Development (FMARD);
- Federal Ministry of Environment (FMEnv);
- Federal Ministry of Health (FMOH);
- Federal Ministry of Industry, Trade and Investment (FMITI);
- Federal Ministry of Science and Technology (FMST);
- Ministry of Foreign Affairs;
- Agricultural Research Council of Nigeria (ARCN);
- National Agency for Food and Drugs Administration and Control (NAFDAC);
- National Agricultural Quarantine Services (NAQS);
- National Agricultural Seed Council of Nigeria (NASCN), Abuja;
- National Biotechnology Development Agency (NABDA);
- National Emergency Management Agency;
- National Planning Commission (NPC);
- National Sugar Development Council (NSDC);
- National Varietal Release Committee;
- Nigeria Customs Services (NCS);
- Nigeria Police Force (NPF);
- Nigeria Security and Civil Defense Corps (NSCDC);
- Standard Organization of Nigeria (SON).

1.6.1.3 National Research Institutes

- Cocoa Research Institute of Nigeria (CRIN), Ibadan;
- Institute of Agricultural Research (IAR), Zaria;
- National Cereals Research Institute (NCRI), Badeggi;
- National Root Crops Research Institute (NRCRI), Umudike;
- National Veterinary Research Institute (NVRI), Jos;

- Nigerian Horticultural Research Institution (NIHORT), Ibadan;
- Nigerian Institute for Oil Research (NIFOR), Benin;
- Nigerian Institute of Trypanosomiasis Research (NITR);
- Raw Materials Research and Development Council (RMRDC);
- Sheda Science and Technology Complex (SHESTCO).

1.6.1.4 Non-Governmental Organizations (NGOs), Civil Society Organizations (CSOs), Professional Organizations, Faith-Based Organizations (FBOs) and Community-Based Organizations (CBOs)

- All Farmers Association of Nigeria (AFAN);
- Nigeria Biological Safety Association (NiBSA);
- Biotechnology Society of Nigeria (BSN);
- Cotton Ginnery Association of Nigeria (CGAN);
- Genetic Society of Nigeria (GSN);
- National Biosafety and Biotechnology Consortium;
- National Cotton Association of Nigeria (NACOTAN);
- Nigerian Society of Microbiologists (NSM);
- Nutrition Society of Nigeria (NSN).

1.6.1.5 Media

- DAAR Communications;
- Daily Trust, Blueprint, Punch and Leadership Newspapers;
- Federal Radio Corporation of Nigeria;
- Nigerian Television Authority, Abuja;
- TVC News;
- News Agency of Nigeria (NAN);
- Voice of Nigeria (VON);
- Armed Forces Radio;
- Daily Trust

(Brief Note from NBMA to the Permanent Federal Ministry of Environment, February 2020).

1.6.2 Memorandum of Understanding (MoU) with Relevant Regulatory Agencies

In furtherance of the Agency's collaborative drive and to enhance synergy among regulatory agencies, the Agency initiated and signed Memoranda of Understanding (MoU) with the following organizations:

- National Agricultural Seed Council (NASC);
- Nigeria Agricultural Quarantine Service (NAQS);
- National Varietal Release Committee (NVRC);
- National Agency for Food and Drug Administration and Control (NAFDAC);
- Standard Organization of Nigeria (SON);
- Nigeria Customs Service (NCS).

1.7 THE ROLE OF NIGERIA IN INTERNATIONAL BIOSAFETY REGULATION

Nigeria has been involved in the following roles in the biosafety regulation:

- Pioneer Chairman, African Union Biosafety Regulators Forum;
- Pioneer Chairman of West African Biosafety Regulation Forum (WABRF);
- Member of West African Biosafety Regulations Committee;
- Vice Chairman, West Africa Integrated Vector Management (WA-IVM) Programme;
- Member of the Ad-Hoc Technical Expert Group on Biosafety Risk Assessment and Risk Management of the Convention on Biological Diversity of the United Nations;
- Member of the Ad-Hoc Technical Expert Group on Social Economic Consideration;
- Nigeria was one of the major voices at the 14th meeting of the Conference of the Parties (COP-14) to the Convention on Biological Diversity, ninth meeting of the Conference of the Parties serving as the meeting of the Parties (COP-MOP-9) to the Cartagena Protocol, third meeting of the conference of the parties serving as the meeting of the Parties (COP-MOP-3) to the Nagoya Protocol held on the 13th–29th of November 2018, in Egypt.

1.8 NEW FOCUS

Administrative focus includes, amongst others:

- Intensive sourcing of befitting Corporate office complex to serve as the Agency's Headquarters and equipment of same with furniture, office equipment, ICT facilities and stationeries;
- Intensive sourcing for land in FCT for the development of a befitting biosafety office complex;
- Acquisition of offices for the new zonal offices (South South, South East, South West, North Central);
- Development of proper transport administration, maintenance and utility services;
- Provision of a backup generator;
- Proper facility management;
- Improved welfare package for staff and security.

Technical focus includes, amongst others:

- Routine Biosafety Surveillance;
- Routine Biosafety Risk Assessment;
- Staff Capacity strengthening (Local Training);
- Sourcing for Foreign Support for Capacity Building (Foreign Training);
- Drafting and development of operational and administrative instruments, i.e. guidelines for the regulation of the emerging modern biotechnology such as gene drive and synthetic biology;

- Development of National Policy on Biosecurity;
- Development of National Biosecurity Emergency Response Strategy;
- Organization of training and enlightenment programmes on emerging modern biotechnology such as gene drive, gene editing, gene sequencing, synthetic biology, next-generation sequencing and biosecurity as well as mainstreaming the same into national agenda;
- Stakeholders Consultative Meeting on Biosecurity;
- Preparation for and attendance of COP 15, COP-MOP 1O, and MOP4;
- Preparation for and attendance of National Council on Environment;
- Upgrade of the ICT facility to facilitate e-governance;
- Integration of NBMA Permit Process into the Nigeria Integrated Customs Information System (NICIS);
- Continuation of the in-house routine and rotational Seminars by members of staff;
- Continuous consultative fora for various importers of GM products;
- Continuous capacity building for staff and other stakeholders;
- Continuous inspection and monitoring of GM facilities as well as survey of GMOs and their products in circulation to ensure biosafety compliance down safety procedures are followed;
- Continuous review of applications for Biosafety Certifications and Permits and granting or refusal of same as necessary;
- Continuous update of NBMA's data repository;
- Establishment of Food Safety Laboratory;
- Establishment of NBMA Secured Database;
- Establishment of physical library and stocking of the Agency's e-library;
- Massive sensitization on the mandate of the Agency and its activities;
- Organization of the 2021 Annual Biosafety Conference;
- Training of the 37 nominated Biosafety State Desk Officers;
- Upgrading of the GM Detection and Analysis Laboratory to bring it to speed with the dynamic nature of modern biotechnology practices;
- Application of data portability, coexistence between GMOs and non-GMOs;
- Training of NBMA staff to assess biosafety risks in agricultural biotechnology;
- Cooperation involving development of standards, formulation and promotion of public policies regarding safe agriculture biotechnology, strengthening regulatory frameworks;
- Optimizing safety and health conditions in biosafety research and development;
- Developing policies to foster biosafety research;
- Promoting exchange of ideas and cooperation in Nigeria and other countries on biosafety and biosecurity issues;
- Biosafety/biosecurity regulation internship.

1.9 REGULATORY CHALLENGES

- **Anti-GMO lobbyist:** Increasing anti-GMO lobbyist and very active anti-GM movement who are aggressively misinforming and misleading stakeholders. The anti-GM movement consists of a number of small organizations that work together to form a united front.

- **Inadequate public awareness:** Resulting in lack of public understanding of the concept of biosafety and its potential benefits to national growth:
- **Third party influences:** Misrepresentation of the Agency's mandate and activities leading to misguided perception of the mandate and activities of the Agency as well as continuous attack by some activists.
- **Poor funding:** Inadequate funding has so far reduced the pace of work of the Agency. Increases and prompt appropriation of funds would improve the overall performance of the Agency,
- **Insufficient vehicles for operational activities:** Hilux pick up vans are needed for fast tracking and surveillance of GM foods and also for ease of transportation to research institutes, universities, CFT and super stores for inspection.
- **Inadequate office space:** Considering the Agency's staff strength, the current office space is inadequate to accommodate the staff.
- **Insufficient office furniture, equipment, ICT facilities and backup power supply:** There is inadequate office furniture such as tables, chairs, computer systems, printers, photocopying machines, UPS and stationeries, etc. as well as inadequate storage facility.
- Inadequate operational uniforms for various sections of the agency.

1.10 CONCLUSION

Nigeria has now joined the League of Nations with sound national biosafety laws and system with hope for higher prospects in biosafety matters globally. Nigeria has put in place an array of regulatory instruments which are products of deliberations by a range of government MDAs and non-governmental agencies, as well as consultations with other relevant stakeholders. These regulatory instruments guide the NBMA in the effective discharge of its duties. Some biosafety decisions reached include approvals for confined field trials and commercial release of GMOs; permits for importation of GMOs for food, feed and for processing; as well as the accreditation and certification of institutes and facilities to carry out modern biotechnology activities.

Nigeria as the Current Chair of the African Union Biosafety Regulators' forum has what it takes to lead the continental forum towards holistic biosafety at various country levels. The National Biosafety System is driven by various experts with various scientific backgrounds with commitment and expertise.

The NBMA has been commended and has various awards for excellent performance of its mandate. Worthy of note is the award of a Certificate of Merit to the NBMA (Figure 1.5), on the occasion of the International Customs Day by the World Customs Organization via the Nigeria Customs Service. The award was presented by the Comptroller-General (CG) of Nigeria Customs Service (NCS), Col. Hameed Ibrahim Ali (Rtd) at the ceremony which took place on the 27th of January 2020 held at NCS HQ, Abuja. The award was in recognition of NBMA's modest achievements.

The Nigerian National Biosafety System is to ensure safety in the application of modern biotechnology towards national economic growth in achieving green economy for sustainable environment.

FIGURE 1.5 Presentation of certification of merit to the NBMA during the international day of customs.

REFERENCES

CBD (1992). *Convention on Biological Diversity*. http://bch.cbd.int.

James, C. (2011).*Global Status of Commercialized Biotech/GM Crops*. Executive Summary, ISAAA Brief No. 42. Ithaca, NY: ISAAA.

NBMA (2017). *National Biosafety Information Manual, 2017*. https://nbma.gov.ng/wp-content/uploads/2021/06/INFORMATION-MANUAL.pdf

National Biosafety Management Agency (2018). *National Biosafety Risk Analysis Framework, 2018*. https://nbma.gov.ng/wp-content/uploads/2021/04/NIGERIA-NATIONAL-BIOSAFETY-RISK-ANALYSIS-FRAMEWORK.pdf

OGTR (2013). Risk Analysis Framework 2013. https://www.ogtr.gov.au/sites/default/files/files/2021-06/risk_analysis_framework_may_2013_0.docx

Oladipo, E., Ogbe, M. G., Molta, N., Ladipo, D., Shingu, G., Maiwada, O., Dore, M. P. O., Olojede, A., Osemeobo, G., Mshelbwala, J., Sarumi, Ojogbo, V., Owolabi, C., Ene-Ita, A., (2001 July) *First National Biodiversity Report*. https://www.cbd.int.doc/world/ng/ng-nr-01-en.pdf.

2 CRISPR/Cas9
Regulatory Considerations to Ensure Safety of Gene-Edited Products in Nigeria

Sylvia Uzochukwu
Federal University

Arinze S. Okoli
University of Tromso

CONTENTS

2.1 INTRODUCTION

Continuous improvement of agricultural products is necessary to offset societal needs for increased productivity and increased quality of plant and animal-derived food products. Man has selected and modified plants and animals from as far back as the Stone Age to achieve improved foods and feeds. Hybrid seeds were responsible to a large extent for the success of the Green Revolution. The adoption of GM crops is seen by some countries as to have made positive economic and production impacts and has also contributed to a more sustainable agriculture (Brookes and Barfoot 2017). However, the technical and socio-economic issues as well as perceived risks to health and environment of the modern genetic modification or transgenics have resulted in very stringent regulations in most countries and regions. Regulation of genetically modified organisms (GMOs) is either process- or product-triggered.

DOI: 10.1201/9781003179177-2

Being process-triggered simply means that if recombinant DNA techniques are used in the development of a product, then it is regulated as a GMO irrespective of the presence or absence of a transgene in the final product. Product-triggered regulations consider novelty of the trait or final product irrespective of the method of development.

2.2 GENETIC MODIFICATION

The Cartegena Protocol defines a GMO or living modified organism (LMO) as ‘any living organism that possesses a novel combination of genetic material obtained through the use of modern biotechnology’, where ‘modern biotechnology’ means ‘the application of:

a. In vitro nucleic acid techniques, including recombinant deoxyribonucleic acid (DNA) and direct injection of nucleic acid into cells or organelles, or
b. Fusion of cells beyond the taxonomic family, that overcomes natural physiological reproductive or recombination barriers and that are not techniques used in traditional breeding and selection’ (Secretariat of the Convention on Biological Diversity, World Trade Centre 2000).

2.2.1 Gene Editing: CRISPR/Cas9 Rules the Waves

Gene editing involves deletion, insertion or modification of specific DNA sequences in a genome by modified enzymes with great precision. Although CRISPR/Cas9 has become synonymous with gene editing, restriction enzymes have initially been applied to edit DNA. However, restriction enzyme-mediated gene editing was neither precise nor targeted. The possibility of targeted gene editing came about with the discovery of zinc-finger nucleases (ZFNs) and transcription activator-like effector nucleases (TALENs). Both ZFN and TALENs involve engineering an enzyme with two domains: DNA-binding domain and a restriction endonuclease domain. The DNA-binding domain targets and binds to specific DNA sequence on the genome, and the endonuclease domain cleaves the DNA at the desired site. Although ZFNs and TALENs can make targeted changes to the genome, they are both inefficient, restricted to certain cell types, very expensive and cumbersome to produce and apply. The CRISPR/Cas9 (Clustered Regularly Interspaced Short Palindromic Repeats (CRISPR)/associated protein 9), which can be applied in all cell types, is based on interactions between DNA, RNA and protein. The system offers a far simpler, cheaper, efficient and easier way of editing genes and has become the popular and preferred way of editing the genome of an organism (Okoli et al. 2018).

2.2.2 Origin of CRISPR/Cas9

The CRISPR/Cas9 is one of the latest introductions to the tools of modern gene (or genome) editing. The origin of the CRISPR/Cas9 system is the immune defence systems of bacteria and archae. In the immune system of bacteria and archae, the CRISPR/Cas system acts in an adaptive manner similar to the adaptive immune system of eukaryotes. The CRISPR/Cas system protects bacteria and archae from invading viruses and plasmids and retains memory of the invasion which is recruited to protect the host in subsequent invasion by the same virus or plasmid (Jinek et al. 2012). In bacteria, these defence systems rely on enzymes (Cas proteins) and small RNAs (CRISPR RNAs-crRNAs) for sequence-specific detection and destruction of the nucleic acids of viruses and plasmids (Jinek et al. 2012).

This technology is now widely explored as a therapeutic strategy against infections (Bikard et al. 2014), various non-malignant and malignant diseases (van Diemen and Lebbink 2017), and in vaccine development and gene therapy (Schwank et al. 2013). In cancer treatment, clinical trials of CRISPR/Cas9-based therapy have been initiated (ClinicalTrials.gov 2017), and several preclinical studies involving CRISPR/Cas9-mediated correction of human genetic diseases are underway. Genome editing in crops can greatly speed up new trait development, and drive improvements in yield and pest and disease resistance as well as biofortification improvement and adaptation to climate change. Genome editing in animals has been used to increase disease resistance, to make livestock better adapted to farming or environmental conditions, to increase fertility and growth, and to improve animal welfare (Ricrock 2019).

2.2.3 Limitations of CRISPR/Cas9

Target specificity and off-target effects: due to the short sequence length of the gRNA (20 nt) and that of the PAM (3 nt), the gRNA-Cas9 complex often binds and mutates at non-specific loci which are similar, but not identical, in homology to target sites. This off-target effect, which is unintended, can result in a number of scenarios, e.g., unexpected secondary effects of the intended modification such as abnormal expression of endogenous genes even when no unexpected sequence changes are detected (Latham et al. 2006). Integration of non-endogenous sequences especially in sites not intended for modification can also result in abnormal expression of endogenous genes (Ladics et al. 2015). The problem of off-target effect presently constitutes one of the important regulatory issues of gene-edited products.

Sub-optimal efficiency in eukaryotic cells: high efficiency (up to 100%) of CRISPR/Cas system has been obtained in bacteria in which the system is naturally expressed (Jiang et al. 2013), but in eukaryotic cells, e.g., plant and human cells, efficiency of the Streptococcus-derived CRISPR/Cas9 varies between 2% and 38% (Mali et al. 2013; Peng et al. 2016). This has been attributed to the large size of the Streptococcus-derived system. However, studies with cells and living organisms have demonstrated up to 100% efficiency with improved CRISPR/Cas9 systems (Moon et al. 2019). Besides not being a natural system in eukaryotes, CRISPR/Cas9 will be expected to be less efficient in editing the double alleles of a densely packed eukaryotic chromosome compared to the simpler haploid bacterial chromosome, but various strategies are being employed to improve efficiencies in non-bacterial systems.

Difficulty in differentiating products modified by CRISPR/Cas9 versus natural modifications: CRISPR/Cas9 system can make mutations in the gene/genome of an organism that involves changes in short length of sequences (1–100 nucleotide bases). Mutations involving short length sequence changes can also be made naturally, thus, creating the problem of being able to differentiate between gene-edited products and products in which the same genetic changes were natural. This is an important issue in the regulation of gene-edited products.

Difficulty in commercializing gene-edited crops that express CRISPR/Cas9: This is primarily due to regulatory constraints in different countries.

2.3 CATEGORIZATION OF GENE-EDITING TECHNIQUES

As the gene-editing technology is rapidly deployed, questions arise as to regulatory approaches for products of the new technology. Three types of genome editing can be distinguished which determine the regulatory approaches to the techniques and their products:

SDN1 (site-directed nuclease 1): This involves the unguided repair of a double strand break (DSB) by non-homologous end joining (NHEJ), which is the natural or default reaction to a DSB. The spontaneous repair of this break could lead to sequence changes or mutations that could result in the silencing of a gene, gene knockout, or a change in the activity of a gene. It is an efficient method, the most widely used, and already has many applications.

SDN2: This involves a template-guided repair of a targeted DSB using a sequence donor, typically short-stranded DNA. The donor carries one or several small mutations flanked by two sequences matching both ends of the DSB, and is thus recognized as a repair template allowing the introduction of the mutation(s) at the target site. The efficiency of the technique is lower than SDN1, but strongly varies according to the species, donor design, and the time and method of delivery and other conditions.

SDN3: This involves a template-guided repair of targeted DSB using a sequence donor that is typically double-stranded DNA containing an entire gene or an even longer genetic element(s). Both ends of the donor are homologous to the DSB ends (usually more than 800 bp each) which therefore recognize the donor as a repair template, allowing the introduction of the gene or genetic element at the target site. Efficiency here is lower than SDN1, but strongly varies according to the species, donor, design, and the time and method of delivery and other conditions (Figure 2.1).

Compared to older techniques, gene editing has lower unintended effects. The specificity of the gRNA makes it possible to identify and minimize potential off-target sites in a sequenced genome, using bioinformatics. However, experimental confirmation of the bioinformatic analysis is necessary. As at present, scientists are working to devise robust methods for experimental identification of off-target effects caused by gene-editing techniques.

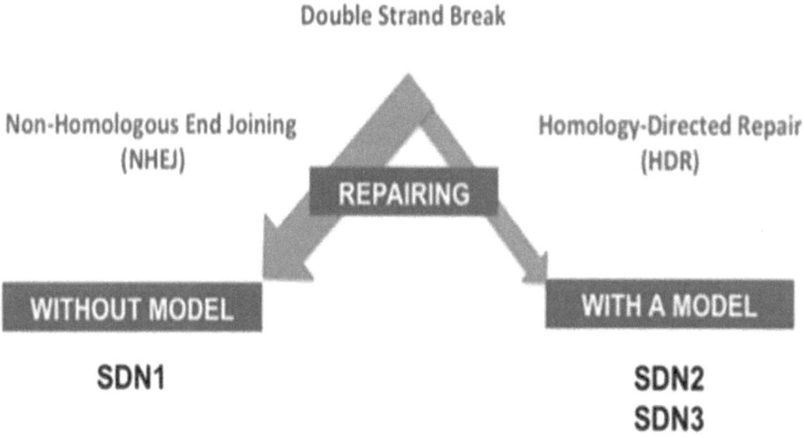

FIGURE 2.1 Mechanisms for double strand break repair during genome editing. The non-homologous end joining (NHEJ) is the main or default double strand break repair mechanism, while homology-directed repair (HDR) occurs at low frequency and requires a template. SDN, Site-Directed Nuclease. (Adapted from Friedrichs et al. 2019.)

2.4 WHY GMOs ARE REGULATED

GMOs are regulated based on the Cartegena Protocol definition of GMO or LMO stated earlier.

The fear is that insertion is random, any gene can be inserted, anywhere, with sometimes unintended results. So there is a global consensus on GMO regulation crafted and overseen by – the Organisation for Economic Cooperation and Development (OECD), the World Health Organization (WHO), and the Food and Agricultural Organization (FAO) of the United Nations. In Nigeria, GMOs are regulated by the National Biosafety Management Agency (NBMA).

GMOs are regulated to ensure they have no adverse effects on the health and safety of humans, animals and the environment. Concerns for the environment include: potential for the GMO to escape and introduce the engineered genes into wild populations, persistence of the gene after the GMO has been harvested, susceptibility of non-target organisms (e.g. insects that are not pests) to the gene product, stability of the gene, loss of biodiversity, and increased use of chemicals. Concerns for human health include concerns for: allergenicity and toxicity of the gene product, gene transfer, especially antibiotic resistance, and out-crossing of genes from GMOs not meant for human consumption. Addressing all these issues is very expensive and time consuming for the developer, with the result that it can take as much as 10 years and millions of dollars, for a GM crop or animal product to get to consumers.

2.5 SHOULD GENE-EDITED PRODUCTS BE REGULATED?

With the advent of genome editing, the question arose globally, as to whether to regulate organisms developed with genome editing techniques using the existing GMO regulatory frameworks, or under new frameworks. As obtains for products of recombinant DNA techniques, central to the need to regulate genome editing techniques is the potential impact of the products on the environment, human and animal health and the society. Hence, an evaluation of potential specific risks associated with genome editing must be considered. For many countries, a regulatory framework for gene-edited crops is yet to be developed. In others, the mechanism for the repair of the DSB determines whether the product can be classified as a GMO or not, using the Cartegena Protocol. The conundrum here is that the Cartegena Protocol is both 'process' and 'product' based. Therefore:

> Using process based, SDN1 and SDN2 both fall within the GMO classification because they were produced using 'modern biotechnology' not minding that they may not possess á novel combination of genetic material′ (for SDN1), or may contain short (less than 100 nucleotide sequence) DNA sequence (for SDN2).
>
> Using product based, it can be argued that the end products of SDN1 and SDN2 do not possess novel genetic material and thus do not fall under the classical GMO definition. This is the argument driving the regulatory approaches of many countries.

2.6 COUNTRY APPROACHES

Presenters at the OECD conference on genome editing in 2018 (Friedrichs et al. 2019) elucidated three major approaches to regulation of gene-edited products by various countries:

1. New regulations for genome-edited products: Argentina is one of the few countries to develop a new set of regulations for gene-edited products.

   Argentina

   Argentina bases its regulation of gene-edited products on the Cartagena Protocol definition of GMO: '...organism that possesses a novel combination of genetic material obtained through ...in vitro recombinant DNA (techniques) and direct injection of nucleic acid into cells.' 'Every gene editing technique involves the use of *in vitro* recombinant DNA techniques (at some stage) and direct injection of nucleic acid into cells.' However, in order to conclude that the organism is GM, it should also have a novel combination of genetic material. Therefore, SDN-1 type gene-edited products are not regulated. For products developed using 'Site-Directed Nucleases type 1 mode' or SDN-1, the change is a point mutation and the final product would not be a GMO. For SDN-2 types the end result will be the same as SDN-1, based on the regulatory criteria, i.e., guided repair with oligonucleotide template. For SDN-3, guided repair with long templates or genes, the final product is a GMO, but there are exceptions, e.g., template from same species. The regulations have a design stage option, where the developers can consult the regulator at the design stage, so that more cost-effective methods can be used. This results in quicker access for the consumer. Another result of this approach is that most gene-edited products are developed by National Small and Medium Enterprises (SMEs) and public research institutions as against big corporations for GMOs. So gene editing results in large numbers of gene-edited products approved per year, thereby fuelling innovation.

2. Review of process-triggered GE/GM regulatory systems: Australia, New Zealand, Europe and India are using a process-driven regulatory trigger to regulate GMOs. They are currently reviewing their regulations to determine if all forms of genome editing fall within their extant GM regulatory frameworks.

European Union

The Court of Justice of the European Union (CJEU) ruled in July 2018 that genome editing techniques fall under the directive EC 2001/18, with no exemption (i.e., they are regarded as GMOs) due to their novelty, with no history of safe use, and will go through the same thorough regulatory process as GMOs. Noteworthy is the fact that history of safe use is the reason for exempting products of conventional mutagenesis from regulation.

India

India is still consulting to determine whether gene-edited products should be regulated under their present regulations for GMOs. However, according to their current definition of gene technology, all new technologies are to be regulated according to existing regulatory framework. That is, all products will undergo the same lengthy regulation as GMOs.

Australia and New Zealand

Australia and New Zealand share a food regulatory system (Figure 2.2). Review of the Gene Technology (GT) Regulations of Australia is still ongoing

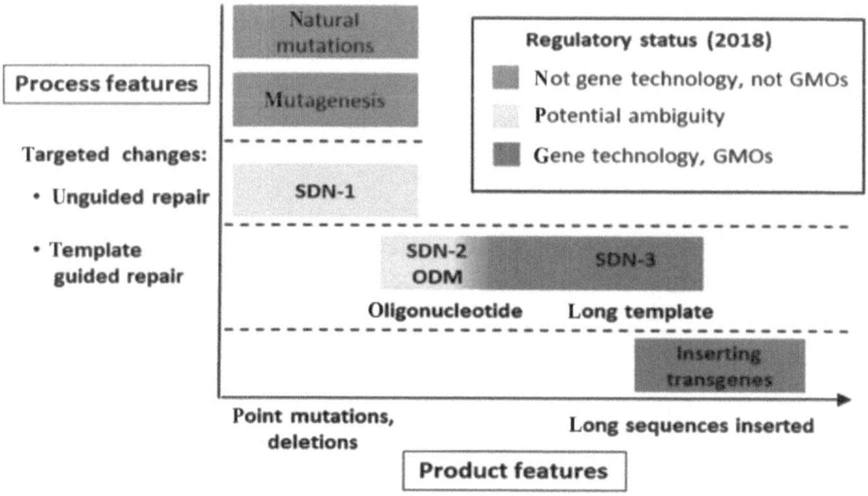

FIGURE 2.2 Regulatory status of products developed with gene-editing techniques in Australia.

and no legislative changes have been made. However, draft amendments proposed would exclude from regulation, organisms produced using site-directed nuclease SDN-1 techniques while organisms produced using oligonucleotide-directed mutagenesis (ODM) or SDN-2, or SDN-3 would continue to be regulated as GMOs. The proposed review would also exclude from regulation, products of transient RNA interference, as well as null segregants (products of crossing of GMOs, that are themselves not transgenic).

Norway

The Norwegian Biotechnology Advisory Board has recommended to the Norwegian government a three-tiered system that recognizes differences in the level of genetic manipulation of the gene-edited product (Norwegian Biotechnology Advisory Board 2018). This is obviously a process-triggered approach.

3. Existing product-triggered regulations: Canada and the United States are using a product-triggered regulatory system for both GMOs and genome-edited products. Relevant novelty of a trait is determined on a case-by-case basis irrespective of the technology used to develop it.

Canada

Canada follows a product-triggered, risk-based regulatory approach to all biotech products Canada requires a pre-market safety assessment for agricultural biotech products, including gene-editing products, only if they are new and can therefore pose a new risk. The country also does not require pre-market risk assessment if the product does not express a novel trait. As in the case of Argentina, proponents are encouraged to contact regulatory authorities early in the product development process to discuss potential regulatory requirements and novelty determination. Still, Canada has identified some policy challenges with gene editing and is holding consultations as to how to solve what they refer to as 'potential problems of regulatory asymmetry' (Friedrichs et al. 2019). This means that SDN-1-type products and most SDN-2 types would not be regulated as they express no new traits.

USA

Like Canada, the United States follows a product-triggered regulation under existing laws, for all biotech products, including gene-edited ones. The USDA declared that most gene-edited agricultural food products do not require regulation. Organisms with the following alterations will not be considered regulated under the USDA-proposed approach: single base pair substitutions, introduction of sequences from sexually compatible plant relatives and complete null segregants.

Japan

Japan has towed the line of the United States and declared that no safety screening would be required for gene-edited food products, as long as the technique used does not leave foreign genes or parts of genes in the target organism. They operate product-triggered regulations.

Nigeria

In August 2019, the NBMA (Amended) Act 2019 was signed into law to expand the oversight of the NBMA, to cover emerging technologies including gene editing, gene drives and synthetic biology.

The Nigerian regulatory framework for GMOs is process triggered. Thus, the approach number 2 above, adopted by the United States and Canada, appear to be out of it. That leaves Nigeria with either revising the existing regulations for GMOs to accommodate genome editing and other emerging biotechnologies, or to develop a completely new set of regulations for crop genome editing as Argentina did (approach number 3 above).

With the relative ease of use and accuracy of the CRISPR/Cas9 system, gene-edited products are flooding the market in Argentina. They are coming from small and medium scale enterprises and academia, in sharp contrast to GMOs that came mainly from big corporations. There is therefore an urgent need to articulate the regulatory approach of Nigeria to genome-edited crops, because, with favourable regulatory conditions, this technology may hold the key to Nigeria's agricultural revolution. Nigeria's population is set to double by 2050, and food production needs to increase at a commensurate rate. Genome editing is one of the technologies that can drive a rapid increase in food production to feed the growing population.

2.7 RECOMMENDATIONS

It is recommended that Nigeria reviews or revises its existing regulations for GMOs to accommodate emerging biotechnologies such as genome editing, gene drives and synthetic biology. Such regulations should be sufficiently 'future proofed' to minimize the need for frequent future reviews and avoid stifling innovation.

It is recommended that in Nigeria's regulatory framework for genome-edited crops, the process should trigger the regulatory process, but the stringency or leniency of the regulation should be based on the nature of the final product.

It is also recommended that Nigeria presents clearly defined protection goals and ecological assessment endpoints – this will help in the country's formulation of a regulatory framework for genome-edited products. Genome editing, if applied in a safe and sustainable manner, can help Nigeria to develop novel agricultural products in a timely manner and greatly boost the country's food security.

In developing new regulations for emerging biotechnologies, efforts should be made to avoid overregulation of gene-edited agricultural products in Nigeria, to ensure the realization and enjoyment of the full potentials of genome-edited agricultural products by Nigerian scientists and consumers, respectively. Thus, a tier-based process in which, as an example, products of SDN1 category are regulated via a more simplified pathway than SDN3 will be helpful.

In all this however, risk analysts and decision makers must be continually concerned with the problem of second order risk, which is the risk of the risk assessment. It takes two forms – the failure to recognize a significant risk, and the failure to recognize that a risk is not significant.

REFERENCES

Bikard, D., Euler, C.W., Jiang, W., Nussenzweig, P.M., Goldberg, G.W., Duportet, X., Fischetti, V.A. and Marraffini, L.A. (2014). Exploiting CRISPR-Cas nucleases to produce sequence-specific antimicrobials. *Nature Biotechnology* **32**(11):1146–1150.

Brookes, G. and Barfoot, P. (2017) Farm income and production impacts of using GM crop technology 1996–2015. *GM Crops Food* **8**(3):156–193.

ClinicalTrials.gov (2017). Available online: http://www.clinicaltrials.gov (accessed 10 October 2017).

Friedrichs, S., Takasu, Y., Kearns, P., Dagallier, B., Oshima, R., Schofield, J. and Morredu, C. (2019). An overview of regulatory approaches to genome editing in agriculture. *Biotechnolology, Research and Innovation* **3**(2):208–220. https://doi.org/10.1016/j.biori.2019.07.001.

Friedrichs, S., Takasu, Y., Kearns, P., Dagallier, B., Oshima, R., Schofield, J., Gill, S.S., Gill, R., Tuteja, R. and Tuteja, N. (2014). Genetic engineering of crops: a ray of hope for enhanced food security. *Plant Signal Behaviour* **9**(3):e28545.

Hsu, P.D., Lander, E. S., and Zhang, F. (2014). Development and applications of CRISPR-Cas9 for genome engineering. *Cell* **157**(6):1262–1278.

Jiang, W., Bikard, D., Cox, D., Zhang, F. and Marraffini, L.A. (2013). RNA-guided editing of bacterial genomes using CRISPR-Cas systems. *Nature Biotechnology* **31**(3):233–239.

Jinek, M., Chylinski, K., Fonfara, I., Hauer, M., Doudna, J.A. and Charpentier, E. (2012). A programmable dual-RNA–guided dna endonuclease in adaptive bacterial immunity. *Science* **337**(6096):816–821.

Jones, H.D. (2019) Future-proofing regulation for rapidly changing biotechnologies. *Transgenic Research* **28**:107–111.

Ladics, G.S., Bartholomaeus, A., Bregitzer, P., Doerrer, N.G., Gray, A., Holzhauser, T., Jordan, M., Keese, P., Kok, E., Macdonald, P., Parrott, W., Privalle, L., Raybould, A., Rhee, S.Y., Rice, E., Romeis, J., Vaughn, J., Wal, J-M., Glenn, K. (2015). Genetic basis and detection of unintended effects in genetically modified crop plants. *Transgenic Research* **24**:587–603.

Latham, J.R., Wilson, A.K., Steinbrecher, R.A. (2006). The mutational consequences of plant transformation. *Journal of Biomedicine and Biotechnology* **2006**:e025376

Mali, P., L. Yang, K.M. Esvelt, J. Aach, M. Guell, J.E. DiCarlo, J.E. Norville and G.M. Church (2013). RNA-guided human genome engineering via Cas9. *Science* **339**(6121):823–826.

Moon, S.B., D.Y. Kim, J.H. Ko and Y.S. Kim (2019). Recent advances in the CRISPR genome editing tool set. *Experimental & Molecular Medicine* **51**(11):1–11.

Norwegian Biotechnology Advisory Board (2018). *A Forward-Looking Regulatory Framework For GMOs.* http://www.bioteknologiradet.no/filarkiv/2018/12/2018-12-03-Sammendrag-genteknologiloven-Bioteknologir%C3%A5det-ENGELSK-for-web.pdf.

Okoli, A., Okeke, M.I., Tryland, M. and Moens, U. (2018). CRISPR/Cas9-advancing orthopoxvirus genome editing for vaccine and vector development. *Viruses* **10**(1):e50.

Paquet, D., D. Kwart, A. Chen, A. Sproul, S. Jacob, S. Teo, K. M. Olsen, A. Gregg, S. Noggle and M. Tessier-Lavigne (2016). Efficient introduction of specific homozygous and heterozygous mutations using CRISPR/Cas9. *Nature* **533**(7601):125–129.

Peng, R., G. Lin and J. Li (2016). Potential pitfalls of CRISPR/Cas9-mediated genome editing. *FEBS Journal* **283**(7):1218–1231.

Ricroch, A. (2019). Global developments of genome editing in agriculture. *Transgenic Research* 28(Suppl 2):45–52. http://dx.doi.org/10.1007/s11248-019-00133-6.

Ricrock, A. (2019). Global developments of genome editing in agriculture. *Transgenic Research* **28**:45–52.

Schwank, G., B.-K. Koo, V. Sasselli, J.F. Dekkers, I. Heo, T. Demircan, N. Sasaki, S. Boymans, E. Cuppen, C.K. van der Ent, E.E.S. Nieuwenhuis, J.M. Beekman and H. Clevers (2013). Functional repair of CFTR by CRISPR/Cas9 in intestinal stem cell organoids of cystic fibrosis patients. *Cell Stem Cell* **13**(6):653–658.

Secretariat of the Convention on Biological Diversity, World Trade Centre (2000). *Cartagena Protocol on Biosafety to the Convention on Biological Diversity*. Montreal: Secretariat to the Convention on Biological Diversity

Stovicek, V., Holkenbrink, C., and Borodina, I. (2017). CRISPR/Cas system for yeast genome engineering: advances and applications. *FEMS Yeast Research* **17**(5):fox030.

Ura, T., Okuda, K. and Shimada, M. (2014). Developments in viral vector-based vaccines. *Vaccines (Basel)* **2**(3):624–641.

van Diemen, F.R. and Lebbink, R.J. (2017). CRISPR/Cas9, a powerful tool to target human herpesviruses. *Cell Microbiology* **19**(2).

van Pijkeren, J.P. and Britton, R.A. (2014). Precision genome engineering in lactic acid bacteria. *Microbial Cell Factories* **13**(Suppl 1):S10.

Wei, Y., Terns, R.M. and Terns, M.P. (2015). Cas9 function and host genome sampling in Type II-A CRISPR-Cas adaptation. *Genes and Development* **29**(4):356–361.

Zhang, Y., W. Qin, X. Lu, J. Xu, H. Huang, H. Bai, S. Li and S. Lin (2017). Programmable base editing of zebrafish genome using a modified CRISPR-Cas9 system. *Nature Communication* **8**(1):118.

3 Regulating Gene-Edited Products in Nigeria

Rufus Ebegba and Joy Onwude
National Biosafety Management Agency (NBMA),
Federal Ministry of Environment

CONTENTS

3.1 INTRODUCTION

The ground-breaking effect of gene editing has been proven on an array of agricultural organisms; the improvement of the efficiency of plant and animal breeding, the introduction of new methods for the management of pests and diseases and the development of new antimicrobials, and attempts to control disease-carrying insects are just a few of its applications (Friedrichs et al., 2019b). This fast-rising deployment of gene editing, however, causes implications on policies pertaining to the technology.

Gene editing which is the modification of the genome within a living cell either by inserting, deleting or replacing one or more segments of DNA is now much easier, faster, cheaper and more versatile than ever; having editing techniques that could allow for new editing applications, ranging from viruses and bacteria, to animals, plants and human beings. According to Nicholas et al. (2017), gene editing is believed to be a game changer for agriculture in the 21st century and beyond as the first wave

DOI: 10.1201/9781003179177-3

of gene-edited products emerge, ranging from soybean with improved oil characteristics to drought-tolerant maize.

Bioethicists, regulators, scientists and other relevant stakeholders have had a series of meetings as regards gene editing, to discuss its applications in agriculture and implications for health and environment. As with any technology, irrespective of its tremendous benefit, there is need for appropriate regulation, for the protection of man and the environment from perceived potential risks that may be associated with it (Ricroch, 2019).

In Nigeria, the National Biosafety Management Agency (NBMA) is charged with the responsibility of providing regulatory framework, institutional and administrative mechanism for safety measures in the application of modern biotechnology and the use and handling of Genetically Modified Organisms (GMOs) in Nigeria (National Biosafety Management Agency, 2015). The Government of Nigeria, in keeping with the dynamics of global technological developments, enacted the NBMA (amended) Act 2019 in August 2019, which expanded the mandate of the NBMA to include emerging biotechnology applications such as gene editing (National Biosafety Management Agency, 2019).

3.2 GENE EDITING AND ITS TECHNIQUES

Gene-editing techniques represent one of the latest innovations in the toolbox of genetic engineering/modification (GE/GM) methods. Some techniques of interest include Transcription Activation-Like Effector Nucleases (TALENs), Zinc-Finger Nucleases (ZFNs), Oligonucleotide Directed Mutagenesis (ODMs) and Clustered Regularly Interspaced Palindromic Repeats (CRISPR). Of these, the most discussed gene-editing system known as CRISPR/Cas9 (Clustered Regularly Interspaced Short Palindromic Repeats, using the CRISPR-associated protein 9) has received widespread application, because it enables the development of low-cost and effective tools for innovation in biomedicine, agriculture, industrial biotechnology and other sectors relating to the bio-economy.

The CRISPR/Cas9 has been moved to the frontline of the Genome Editing Technologies for plants because its procedures are highly precise and simple. In addition, it possesses little intellectual property restriction, as well as being quicker and easier to use. The principle of gene-editing technology is grounded on chimeric proteins made up of sequence-specific DNA-binding domains and non-specific DNA cleavage nucleases.

Gene-editing techniques are categorized into site-directed nuclease (SDN)-1, SDN-2, and SDN-3. SDN-1 involves the unguided repair of a targeted double-strand break (DSB) by the mechanism called non-homologous end joining. The spontaneous repair of this break can lead to a mutation causing gene silencing, gene knock-out or a change in the activity of a gene. SDN-2 involves a template-guided repair of a targeted DSB using a sequence donor, typically short single-stranded DNA. The donor carries one or several small mutations flanked by two sequences matching both ends of the DSB, and is thus recognized as a repair template, allowing the introduction of the mutation(s) at the target site. SDN-3 involves a template-guided repair of a targeted DSB using a sequence donor, typically double-stranded DNA containing an entire gene or an even longer genetic element. Both ends of the donor are homologous to the DSB ends, which therefore recognize the donor as a repair template, allowing the introduction of the gene or genetic element(s) at the target site (Ricroch, 2019; Friedrichs et al., 2019a). Examples of plants in which gene-editing studies are being carried out include barley, maize, wheat, soybean, sorghum and rice (Waquar et al., 2020).

3.3 GLOBAL REGULATORY APPROACHES TO GENE EDITING

Regulatory systems have been classified into two major types: process-triggered genetic engineering/modification (GE/GM) regulatory systems and product-triggered regulatory system. Australia, New Zealand, Europe and India are using a process-driven regulatory trigger to regulate GE/GM organisms; these jurisdictions are reported to be currently reviewing the scope of their regulatory definitions, in order to clarify, if all forms of gene editing fall under their respective existing GE/GM regulatory framework. On the other hand, Canada and the United States are regulating GE/GM and gene-editing products according to a product-trigger under which the relevant novelty of the trait in question is considered on a case-by-case basis, irrespective of the technology used to develop it. However, Argentina is reported to have introduced a new, customized regulatory resolution on New (Plant) Breeding Techniques (N(P)BTs) in 2015, making it one of the first countries to have passed a regulation on this novel set of techniques, covering the sub-category of Gene Editing in their course (Friedrichs et al., 2019c).

Australia has launched a public consultation on GMO regulation amendments (Office of the Gene Technology Regulator, OGTR). As a consequence of the proposed changes, some cases of gene editing may be exempted from regulation. However, the scope of potentially exempted products in Australia would be quite narrow compared to the approaches of the other countries that have made regulatory decisions until now. In the United States, current decisions on gene-edited plants have been based on the Plant Protection Act, as enforced by the United States Department of Agriculture (USDA). The US Coordinated Framework for Biotechnology makes no special provisions for gene-edited crops. As for any biotechnology-derived plant, if the gene-edited crop poses a plant pest risk, expresses a pesticide trait or poses food safety risks different from other plants produced through traditional plant breeding, then it is subject to regulatory considerations (by United States Department of Agriculture, USDA; Environmental Protection Agency, EPA; and Food and Drug Administration, FDA, respectively); otherwise, the product can freely enter market channels.

3.4 GENE EDITING IN NIGERIA

The National Biosafety Management Act 2015 (as amended), in section 25(a), states "*No person, institution or body shall carryout gene drive, gene editing and synthetic biology except with the approval of the Agency*". The regulation of gene editing in Nigeria as well as information contained in the National Guidelines on Gene Editing is limited to gene-editing techniques as it relates to plants, animals and microorganisms – process and product – as it applies throughout the Federal Republic of Nigeria.

3.5 REGULATION OF GENE EDITING IN NIGERIA

The NBMA being empowered by the National Biosafety Management Agency Act 2015 (as amended) aiming amongst others to identify and develop functional strategies to facilitate its implementation has led to the production of the National Guidelines on Gene Editing. The technical document provides pertinent information to applicants, stakeholders and the general public on the NBMA's statutory biosafety oversight on gene-editing activities. It explains processes and procedures of the NBMA in dealing with applications for gene editing and highlights the requirements of an applicant (National Guidelines on Gene Editing, 2020).

As explained in the previous sections, rules that determine whether or not an organism falls under a special GMO biosafety regulatory regime differ from one country to the other. Quite often, their factors for regulatory inclusion are based on either the processes used to obtain them or the product characteristics. However, in Nigeria, if the gene-editing process or the gene-edited product does not lead to or have a new combination of genetic material (e.g. does not use a transgene or uses a transgene which is removed in the final product), a non-GM regulatory classification is applied: in this case, basic information on the overall breeding process, genetic changes, traits, bred-out of helper transgenes, etc. will be required to be submitted to NBMA prior to release of the product.

If, on the other hand, the gene-editing process requires the use of recombinant DNA sequences or the gene-edited product has a new combination of genetic material (e.g. uses a transgene that remains in the final product), the regulatory classification stipulates that the final product falls under GM regulation (See Figure 3.1).

3.5.1 Nigeria's Provision for Application for Dealings with Gene-Edited Products

The NBMA has set out mandatory steps for any person, group of persons, institution or company intending to have dealings with gene editing for any purpose (containment, confined field trial, multi-locational trial, commercial/general release, import, export, food/feed/for processing, transport and any other activity) to take. This includes obtaining and completion of the application form (from the NBMA website – www.nbma.gov.ng or offices) and submission of form in a dossier, alongside all relevant documents, to the NBMA. The NBMA then carries on with a series of processes which include: acknowledgement of application by the Agency; check for completeness of Application Dossier/internal review of the application; request for additional information by the Agency, if need be; and conveyance of decision of the internal review. The decision of the internal regulatory review will indicate if a Biosafety Authorisation/Clearance will be issued on a case-by-case basis (where the process or the product is found not to contain any foreign DNA) or whether the review of the application will continue further in situation where the process or the product is classified as GMO.

In the situation where the application assessed falls under GM category and must be regulated as such, the applicant will proceed to pay the required non-refundable application fee. Following payment by an applicant, the Agency will display the

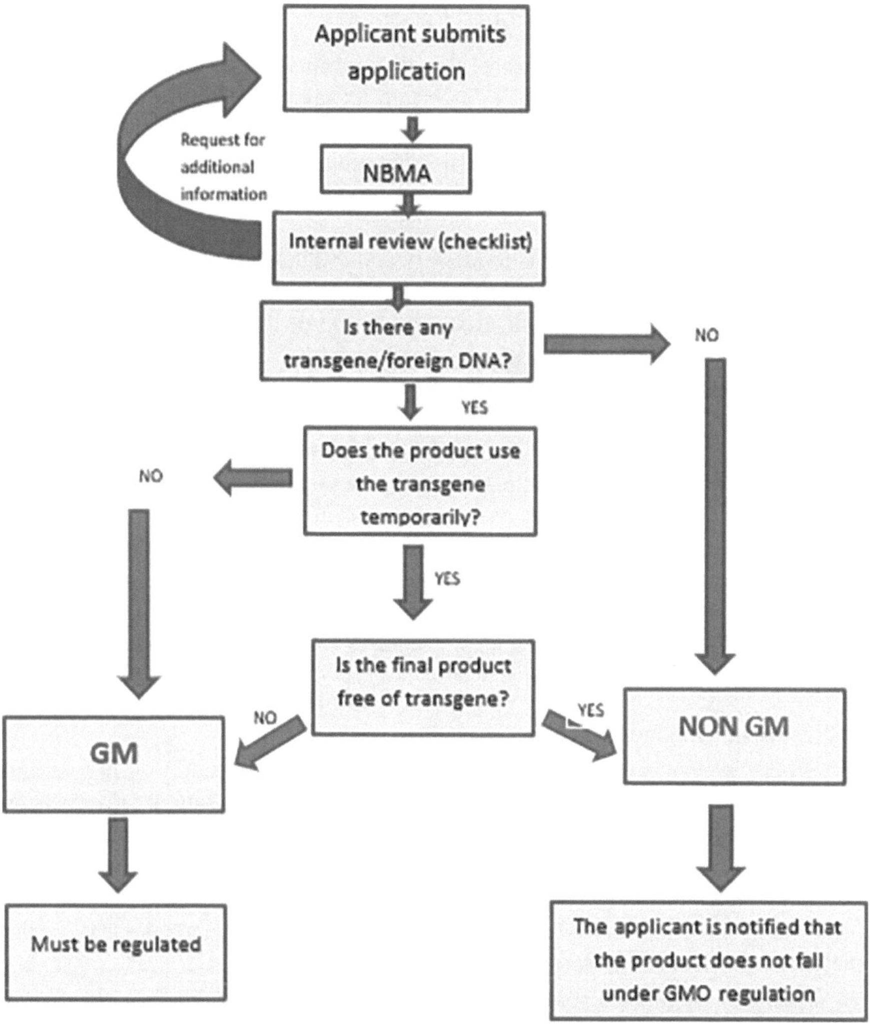

FIGURE 3.1 Nigeria's process map towards decision making.

Application Dossier to the general public, the notice of display of the application will be published in at least two national dailies and one local newspaper, the Biosafety Clearing House, Agency website or such other news media as the Agency may from time to time determine.

The National Biosafety Committee and National Biosafety Technical Sub-Committees would thereafter be constituted for in-depth review of application. Subsequently, the Agency shall issue Permit or reject application on the basis of recommendation by NBC and the decision of the DG/CEO. This decision will be forwarded to relevant regulatory Agencies and also be made available in the Biosafety Clearing House (BCH), NBMA website and other news media (National Guidelines on Gene Editing, 2020).

3.5.2 Risk Analysis of Gene-Edited Product

Prior to approval of any activity concerning gene editing to be regulated, a comprehensive risk analysis is carried out to ascertain its safety to human health and the environment. This risk analysis process consists of three interconnected elements: risk assessment, risk management and risk communication.

3.5.2.1 Risk Assessment

The objective of risk assessment is to identify and evaluate potential adverse effects of the gene-edited product, direct or indirect, immediate or delayed, on human and animal health and the environment. Risk assessment of the gene-edited product is carried out on a case-by-case and step-by-step basis.

Risk assessments takes amongst others the following parameters into consideration: the description of the modifications made using gene technology; the function of the genetic modifications and the new insert including any marker gene(s); purpose of the modification and intended use in relation to need and benefit; method of modification; number of inserts/deletions/replacement; position in the genome; and its or their structures (e.g., the copy number whether in random or other types of repeats); products of the transferred genes; genetic changes; traits; bred-out of helper transgenes; level of expression and methods for measuring expression; stability of the introduced genes in terms of expressions, structures and sites of integration (National Biosafety Management Act, 2015).

3.5.2.2 Risk Management

The purpose of risk management is to protect the health and safety of people and the environment by controlling or mitigating risk. For gene editing, it encompasses preparation of a risk management plan which includes training on general risk management measures, evaluation and mitigation of risks, and proposed licence conditions as well as monitoring and review which details measures to assess effectiveness of all steps in risk analysis including post release review of commercial release of the gene-edited products of its derivatives thereof.

3.5.2.3 Risk Communication

Good risk communication promotes a clear understanding of all aspects of risk and to provide information about risk to help people make decisions, to minimize conflicts, to improve understanding of perceptions and positions, and to achieve equitable outcomes. Appropriate communication channels where information concerning gene editing is passed includes the Agency's website – www.nbma.gov.ng – and the BCH.

3.6 UNINTENDED EFFECTS

Gene-editing techniques involving CRISPR-Cas9 editing technology has been of tremendous benefit for scientists trying to know and appreciate the role of genes in disease. This technique also has resulted in elevated hopes for more powerful gene therapies that can delete or repair flawed genes (Schaefer et al., 2017). But even though CRISPR can precisely target specific stretches of DNA, it sometimes hits other parts of the genome. In a study of the unexpected mutations after CRISPR Cas9 editing in vivo by Schaefer et al. (2017), it was concluded that gene-editing technology can introduce hundreds of unintended mutations into the genome (Schaefer et al., 2017). Results of some studies of unintended effects that have been carried out include exosome-mediated horizontal gene transfer which occurred in DSB repair during genome editing (Ono et al., 2019); large deletions and complex rearrangements formed during repair of DSBs induced by CRISPR-Cas9 (Kosicki et al., 2018); large deletions of whole exons induced by a single gRNA causing exon skipping in cell lines (Sharpe and Cooper, 2017); and recombination between homologous chromosome arms triggered by CRISPR-induced DSBs (Bruner et al., 2019). Kawall in 2019 concluded that new genome editing technologies, such as CRISPR/Cas9, are now making the entire genome accessible for any desired change by the researcher and breeder. "These new techniques circumvent mechanisms that protect certain areas of the genome by targeting nucleases to specific genomic regions, thereby increasing the probability of the induction of genomic alterations" (Kawall, 2019).

The stability and off-target sites of CRISPR/Cas9 editing in some plants and cereals like rice are largely unknown, although there are several reports on the immediate generations after transformation (Zhang et al., 2014; Zhou et al., 2014); however, the multi-generation analysis by Feng et al. revealed the inheritance, specificity and patterns of CRISPR/Cas-induced gene modifications in *Arabidopsis*. CRISPR/Cas9 was seen to be very effective in targeted gene mutagenesis (dominated by small indels), resulting homozygous lines were stable for the mutations in subsequent generations, and off-target mutations were rare (Feng et al., 2014). Gene editing using CRISPR/Cas9 system may be not as precise as expected in rice and some other plants, as such early and accurate molecular characterization and screening must be carried out for generations before transitioning of the CRISPR/Cas9 system from laboratory to field (Biswas et al., 2020).

3.7 ETHICAL CONCERNS OF GENE EDITING

Worldwide attention has been particularly focused on the issues surrounding clinical application of gene editing to humans. Most of the ethical discussions related to gene-editing centre around human germline because editing changes made in the germline would be passed down to future generations (National Institutes of Health, 2017).

Given that CRISPR technology has the potential to make gene editing more accurate and even "easy" in comparison to older technologies, the debate about such editing has now regained tremendous attention inasmuch as it is a relatively old subject. Some bioethicists and researchers generally believe that human genome editing for reproductive purposes should not be attempted at this time, but that studies that would make gene therapy safe and effective should continue (National Academies of Sciences, 2017; The Hinxton Group, 2015). As of 2014, there were about 40 countries that discouraged or banned research on germline editing, including 15 nations in Western Europe, because of ethical and safety concerns (Araki and Ishii, 2014). Although Lanphier et al. in 2015 concluded that there may never be a time when genome editing in embryos will offer a benefit greater than that of existing technologies, such as preimplantation genetic diagnosis (PGD) and in-vitro fertilization

(IVF), other scientists and bioethicists acknowledge that in some cases, germline editing can address needs not met by PGD, this includes cases of polygenic disorders, which are influenced by more than one gene (Hampton, 2016; Savulescu et al., 2015).

In Nigeria, however, current application of gene editing is focused on improved agriculture. The clinical application of this technique is still a distant prospect.

3.8 SOCIO-ECONOMIC CONSIDERATIONS

For any gene-edited product that contains a novel DNA, it is considered as a GMO and as such, socio-economic considerations for GMOs shall apply. The socio-economic considerations focus on those with impacts on the conservation and sustainable use of biological diversity, especially with regard to the value of biological diversity to indigenous and local communities.

The National Biosafety Management Act (2015, as amended) has been sculptured to ensure that production and use of genetically modified organisms take place in an ethical and socially justifiable way, in harmony with the principle of sustainable development and without detrimental effects on health and environment. The NBMA ensures that approval is not being given to the placing on the market of a gene-edited product except it is considered and duly determined by the Agency to: benefit Nigeria without causing significant risk to human health, biological diversity and in general the environment; contribute to sustainable development; not have adverse socio-economic impacts; and accord with the ethical values and concerns of communities and does not undermine indigenous knowledge and technologies.

Prior to placing on the market of gene-edited products; an applicant, following the directive of the Agency carries practical steps to ensure that socio-economic factors are considered. Some of which include: conducting a comprehensive socio-economic assessment of each product on each application, on a case-by-case basis; submission of socio-economic literature review and analysed data of the baseline survey done in the vicinity; conduction of socio-economic impact assessment within the vicinity amongst relevant stakeholders (farmers, household and community; industry and markets; consumers; and traders).

The National Biosafety Committee and the National Biosafety Technical Sub-Committee play a vital role as it evaluates the inputs and analysed data of the socio-economic consideration alongside the risk assessment report and the risk management plan, and present its findings to the Agency for consideration.

3.9 PACKAGING, IDENTIFICATION AND TRANSPORT OF GENE-EDITED PRODUCTS

The NBMA in the National Guidelines on Gene Editing, 2020, has set out procedures for packaging, labelling and identification as well as transport of gene-edited products. Identification of gene-edited products will follow the labelling guidelines recommended by the Agency in section (3) of the first schedule of the NBMA Act 2015. NBMA ensures that a gene-edited product classified as GMO or its derived product(s) is clearly identified and labelled, the unique identifier suitable for each gene-edited product is included in the labelling and the language on the label of gene-edited products is written in English or has an English translation written on the label or package insert in cases where it is labelled in another language. Any gene-edited product that is considered to be genetically modified and is to be imported into Nigeria, exported from Nigeria or transported within or across Nigeria will be accompanied by the appropriate documentations as stated in the NBMA Regulations 2017.

3.10 CONCLUSION

Gene editing through the use of CRISPR/Cas9 systems open up a new opportunity for quick and precise modification of crops, for example, to boost productivity (Jones 2015; Li et al., 2017), to protect against pests and diseases (Wang et al., 2014) and to enhance nutrient content (Liang et al., 2014). The application of CRISPR/Cas9 in crop improvement has, however, raised increasing concerns about the safety of its products. Appropriate research and development investments should be encouraged so as to develop greater competencies to determine safety in CRISPR and related fields to confront challenges in the grand scheme of things.

Nigeria is taking leadership in formulating adequate policies and developing guidelines and regulations through the NBMA. The NBMA is well positioned to effectively manage and ensure safety in the application of modern biotechnology and use of genetically modified organisms in Nigeria, with a view to avert any adverse impacts on human health and the environment. Obtaining regulatory approvals, among other factors, is central to the application of modern biotechnology and the deployment of its products for socio-economic benefit.

For the full discharge of the mandate of the NBMA, the support of all Nigerian citizens is requested. All are advised to play by the rules as breach of the National Biosafety Management Agency Act 2015 (as amended), National Guidelines on Gene Editing 2020 and other pursuant regulations and guidelines have severe consequences. The provisions on offences, penalties and enforcement in Part IX of the NBMA Act 2015 shall also apply to gene-editing activities.

REFERENCES

Araki, M. and Ishii, T. (2014). International regulatory landscape and integration of corrective genome editing into in vitro fertilization. *Reproductive Biology and Endocrinology*, 12(108): 1–13.

Biswas, S., Tian, J., Li, R., Chen, X., Luo, Z., Chen, M., Zhao, X., Zhang, D., Persson, S., Yuan, Z. and Shi, J. (2020). Investigation of CRISPR/Cas9-induced SD1 rice mutants highlights the importance of molecular characterization in plant molecular breeding. *Journal of Genetics and Genomics*, 47(5): 273–280.

Bruner, E., Yagi, R., Debrunner, M., Beck-Schneider, D., Burger, A., Escher, E., Mosimann, C., Hausmann, G. and Basler, K. (2019). CRISPR-induced double-strand breaks trigger recombination between homologous chromosome arms. *Life Science Alliance*, 2(3).

Feng, Z., Mao, Y., Xu, N., Zhang, b., Wei, P., Yang, D. L., Wang, Z., Zhang, Z., Zheng, R., Yang, L., Zeng, L., Liu, X., and Zhu, J. K. (2014). Multigeneration analysis reveals the inheritance, specificity, and patterns of CRISPR/Cas-induced gene modifications in *Arabidopsis*, *Proceedings of the National Academy of Science of the United States of America*, 111: 4632–4637.

Friedrichs, S., Takasu, Y., Kearns, P., Dagallier, B., Oshima, R. and Schofield, J (2019a). Meeting report of the OECD conference on genome editing: applications in agriculture – implications for health, environment and regulation. *Transgenic Research*, 28: 419–463.

Friedrichs, S., Takasu, Y., Kearns, P., Dagallier, B., Oshima, R., Schofield, J. and Catherine M. (2019b). An overview of regulatory approaches to genome editing in agriculture. *Biotechnology Research and Innovation*, 3(2): 208–220.

Friedrichs, S., Takasu, Y., Kearns, P., Dagallier, B., Oshima, R. and Schofield, J. (2019c). Policy considerations regarding genome editing. *Trends in Biotechnology*, 37(10): 1092–1032.

Hampton, T. (2016). Ethical and societal questions loom large as gene editing moves closer to the clinic. *JAMA*, 315(6): 546–548.

Jones, H. D. (2015). Regulatory uncertainty over genome editing. *Native Plants*, 1: 1–3.

Kawall K (2019). New possibilities on the horizon: genome editing makes the whole genome accessible for changes. *Frontiers in Plant Science*, 10: 525.

Kosicki, M., Tomberg, K. and Bradley, A. (2018). Repair of double-strand breaks induced by CRISPR-Cas9 leads to large deletions and complex rearrangements. *Nature Biotechnology*, 36: 765–771.

Lanphier, E., Urnov, F., Haecker, S. E., Werner, M. and Smolenski, J. (2015). Don't edit the human germ line. *Nature News*, 519(7544).

Li, X., Zhou, W., Ren, Y., Tian, X., Lv, T., Wang, Z., Fang, J., Chu, C., Yang, J. and Bu, Q. (2017). High-efficiency breeding of early-maturing rice cultivars via CRISPR/Cas9-mediated genome editing. *Journal of Genetics and Genomics*, 44: 175–178.

Liang, Z., Zhang, K., Chen, K. and Gao, C. (2014). Targeted mutagenesis in *Zea mays* using TALENs and the CRISPR/Cas system. *Journal of Genetics and Genomics*, 41: 63–68.

National Academies of Sciences (2017). *Human Genome Editing: Science, Ethics, and Governance*. Washington, DC: The National Academies Press.

National Institutes of Health, National Human Genome Research Institute (2017). *What are the Ethical Concerns of Genome Editing?* Retrieved from https://www.genome.gov/about-genomics/policy-issues/Genome-Editing/ethical-concerns.

NBMA (2015). *National Biosafety Management Agency Act, as Amended*. https://nbma.gov.ng/wp-content/uploads/2021/04/Amendment-of-NBMA-Act-2019.pdf

NBMA (2017). *National Biosafety Information Manual, 2017,*. https://nbma.gov.ng/wp-content/uploads/2021/06/INFORMATION-MANUAL.pdf

NBMA (2017). *National Biosafety Risk Analysis Framework, 2019,*. https://nbma.gov.ng/wp-content/uploads/2021/04/NIGERIA-NATIONAL-BIOSAFETY-RISK-ANALYSIS-FRAMEWORK.pdf

NBMA (2020). *National Guidelines on Gene Editing,*. https://nbma.gov.ng/wp-content/uploads/2021/04/NBMA-GENE-EDITING-GUIDELINE.pdf

Nicholas J.B., Javier G. and Daniel F.V. (2017). Genome engineering and agriculture: opportunities and challenges. *Progress in Molecular Biology and Translational Science*, 149: 1–26.

Ono, R., Yashuhiko, Y., Aisaki, K., Kitajima, S., Kanno, J. and Hirabayashi, Y. (2019). Exosome-mediated horizontal gene transfer occurs in double-strand break repair during genome editing. *Communications Biology*, 2: 57.

Ricroch, A. (2019). Global developments of genome editing in agriculture. *Transgenic Research*, 10.1007/s11248-019-00133–6.

Savulescu, J., Pugh, J., Douglas, T. and Gyngell, C. (2015). The moral imperative to continue gene editing research on human embryos. *Protein Cell*, 6(7): 476–479.

Schaefer, K. A., Wu, W., Colgan, D. F, Tsang, S. H., Bassuk, A. G., and Mahajan, V. B. (2017). Unexpected mutations after CRISPR Cas 9 in vivo. *Nature Methods*, 14: 547–548.

Sharpe, J. J. and Cooper, T. A. (2017). Unexpected consequences: exon skipping caused by CRISPR-generated mutations. *Genome Biology*, 18(1): 109.

The Hinxton Group (2015). *Statement on Genome Editing Technologies and Human Germline Genetic Modification.* Retrieved from http://www.hinxtongroup.org/Hinxton2015_Statement.pdf.

Wang, Y., Cheng, X., Shan, Q., Zhang, Y., Liu, J., Gao, C., Qiu, J. L. (2014). Simultaneous editing of three homoeoalleles in hexaploid bread wheat confers heritable resistance to powdery mildew. *Nature Biotechnology*, 32: 947–951.

Waquar, A. A., Sonali U. C., Vacha, B., Altafhusain, B. N., Sanskriti, V., Jawahar, L. K., Humira, S. and Rupesh, D. (2020). Genome editing in cereals: approaches, applications and challenges. *International Journal of Molecular Sciences*, 21: 4040.

Zhang, H., Zhang, J., Wei, P., Zhang, B., Gou, F., Feng, Z., Mao, Y., Yang, L., Zhang, H., Xu, N. and Zhu, J. K. (2014). The CRISPR/Cas9 system produces specific and homozygous targeted gene editing in rice in one generation. *Plant Biotechnology Journal*, 12: 797–807.

Zhou, H., Liu, B., Weeks, D. P., Spalding, M. H. and Yang, B. (2014). Large chromosomal deletions and heritable small genetic changes induced by CRISPR/Cas9 in rice. *Nucleic Acids Research*, 42: 10903–10914.

4 Emerging Technologies
Can African Regulators Take a Lead on Transgenic Mosquito Regulation?

Rufus Ebegba and Joy Onwude
National Biosafety Management Agency (NBMA),
Federal Ministry of Environment

CONTENTS

4.1 INTRODUCTION

Technology has always played a key role in different sectors, and the environment, health care and agricultural industry are moving towards a new frontier of human well-being with the advances in technology. Artificial intelligence, Internet of Things (IoT), cybersecurity, robotics/drones, genetic editing, gene drives, sensors and vertical farming are some of the prime trends (Bongomin et al., 2020). This trend is expected to grow in coming years, as many technologies in various sectors are being introduced. Modern biotechnology, for example, has birthed inventions that wasn't even thought possible in the early 1900s. The specificity of Clustered Regularly Interspaced Short Palindromic Repeats and CRISPR associated proteins-9 (CRISPR/Cas-9) is a remarkable breakthrough in the agricultural sector and has been found to be a useful tool in genome editing amongst others, and there is, however, a need for adequate regulation of all technologies (Ricroch, 2019).

In Nigeria, the National Biosafety Management Agency (NBMA) is charged with the responsibility of providing regulatory framework, institutional and administrative mechanism for safety measures in the application of modern biotechnology and the use and handling of GMOs in Nigeria (National Biosafety Management Agency, 2015).

DOI: 10.1201/9781003179177-4

This is with a view to preventing any adverse effect on human health, animal, plant and the environment. Furthermore, the Government of Nigeria, in keeping with the dynamics of global technological developments, endorsed the NBMA (amended) Act 2019 in August 2019, which expanded the mandate of NBMA to include Biosecurity as well as emerging biotechnology applications: gene drive, synthetic biology and gene editing (National Biosafety Management Agency, 2019). Some African countries also have a National Biosafety Authority in place (Morris, 2014).

4.2 NEED FOR TRANSGENIC MOSQUITOES

The African continent is the most affected by malaria, with 90% of the world's 216 million cases in 2016 recorded in sub-Saharan Africa (WHO, 2017). The disease causes extensive economic losses, with recent estimates suggesting that countries which are severely affected by malaria have up to five times lower Gross Domestic Product (GDP) than those without malaria (Jobin, 2014). There has been significant progress against the disease in the past decade and a half, mostly due to the scale-up of long-lasting insecticide-treated bed nets (LLINs), indoor residual spraying (IRS), and artemisinin-based combination therapies (ACTs). Between the years 2000 and 2015, an estimated 663 million malaria cases were prevented, with insecticidal bed nets and IRS together contributing 78% towards these gains (Bhatt et al., 2015). Despite these successes, residual levels of malaria transmission still persist in several areas across the continent, even in places where coverage with existing interventions already exceeds 80%. Complete elimination of malaria remains a challenging goal across most sub-Saharan Africa settings due to factors such as insecticide and drug resistance, low funding which affects compliance to current best practices, sub-optimal strategies and poor living conditions. Other factors include human activities and mosquito behavioural adaptations. For example, although LLINs and IRS have effectively tackled mosquitoes that bite humans inside houses, they may not adequately control mosquitoes that bite outdoors or early in the evenings (Durnez and Coosemans, 2013; Elliott, 1972; Pates and Curtis, 2005). Experts now believe that without additional complementary interventions, the African Union target for malaria elimination by 2030 will not be achieved, hence the need to explore emerging technologies such as gene-drive technology for malaria control and elimination.

4.3 GENE DRIVES

Gene drives are engineered snippets of DNA that can be introduced into an organism's genome to significantly increase the chance that a desired genetic trait will spread through a population faster than would normally happen through sexual reproduction (Friedman et al., 2020). Gene-drive techniques have already been demonstrated in laboratories to effectively alter Anopheles mosquito populations so that they can no longer transmit malaria parasites (Gantz et al., 2015), and also to introduce lethal gene sequences that suppress and rapidly crash entire mosquito populations in laboratories (Hammond et al., 2016). Mathematical evaluations of these approaches indicate that, if combined with existing interventions like LLINs, these technologies can effectively eliminate malaria in several African settings within a few years after the initial releases of even small numbers of modified mosquitoes (Eckhoff et al., 2016). It is particularly important to emphasize that the use of transgenic mosquitoes of any kind, as with any other vector control tool, should be considered as just one component of an integrated approach, rather than as stand-alone technology (Marshall and Taylor, 2009).

Suppression drives and replacement drives are amongst the general classes of traits or effectors being used to control mosquito pests. Suppression drives employ genetic modifications that reduce or eliminate pest populations, in this case, the number of disease-carrying mosquitoes is reduced (Friedman et al., 2020). Replacement drives on the other hand modify pest populations so that they are no longer disease vectors or pests, and this is exemplified in a case whereby a gene that renders mosquitoes unable to transmit malaria is added. The number of mosquitoes remains the same; only the number of malaria-causing parasites is reduced (Friedman et al., 2020).

The advent of synthetic CRISPR/Cas9-based gene-drive systems allows very specific gene editing in the mosquito, as well as validation of genes expected to be essential in processes of interest for malaria control, such as reproductive capacity, mate-seeking behaviour, immunity against parasites and sex determination (NEPAD, 2018).

4.4 TRANSGENIC MOSQUITOES REGULATION IN AFRICA

Currently in Africa, there is an active development of gene-drive technologies to suppress or alter malaria-transmitting mosquitoes with potential field applications. The WHO Vector Control Advisory Group (VCAG) have received submissions from about two research groups for evaluation on an initial work on gene drives (WHO Vector Control Advisory Group, 2017); however, several other groups are also working on gene drives for malaria control. Gene-drive mosquitoes have been produced in laboratories (Gantz et al., 2015; Hammond et al., 2016) and are expected to be eventually tested in the field in Africa. A small number of African institutions and scientists are also engaging actively in these developments and regulatory processes (see Figure 4.1).

Burkina Faso scientists have obtained authorization from their National Biosafety Agency to import "sterile male" transgenic mosquitoes without any gene drive, for purposes of capacity building, public engagement and strengthening of internal regulatory processes. The eggs have been received and the mosquitoes are being maintained in a contained facility at Institut de Recherche en Sciences de la Santé (IRSS) in Bobo Dioulasso, southwestern Burkina Faso (NEPAD, 2018). Burkina Faso has shown preparedness by the establishment of fully functional containment facilities for transgenic mosquitoes while Mali and Uganda are developing theirs. The Mali team has recently submitted its application for contained trials to the regulatory body and are awaiting their feedback (Glover et al., 2018).

In Nigeria, the President of the Federal Republic endorsed the NBMA (amended) Act 2019 in August 2019, which expanded the mandate of NBMA to include gene-drive as well as other emerging biotechnology applications (NBMA, 2019). Guidelines and regulations to be developed for this purpose are underway.

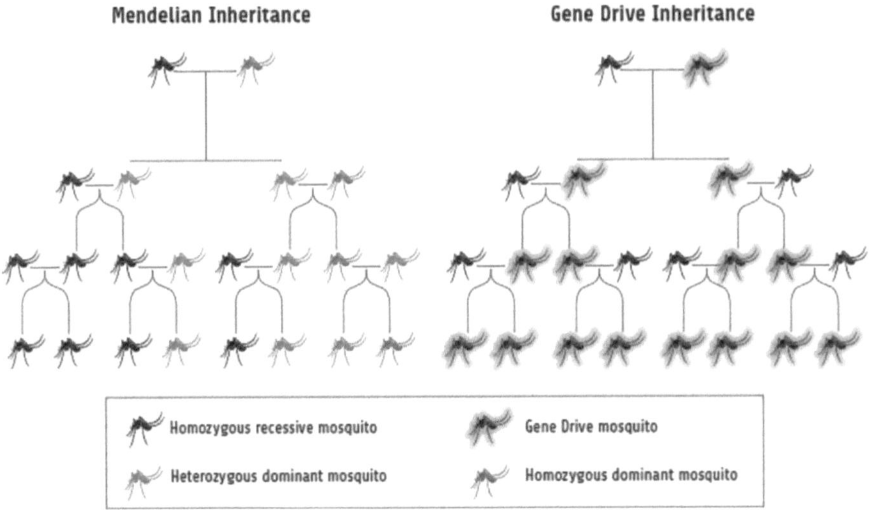

FIGURE 4.1 Illustration of the gene-drive concept, compared to normal Mendelian inheritance (NEPAD 2018). Gene drives for malaria control and elimination in Nigeria.

As for Uganda, a contained facility laboratory will be constructed and the team will proceed with the application for contained trials once the facility is ready. All this progress will ensure that other countries that come on board can circumvent any challenges and benefit from the experiences of Burkina Faso, Uganda and Mali. Most of the work, however, is still in the early stages and will require further validation and field-testing. It is likely that the technology will take several years (estimated to be 10 years) before actual field-deployable gene-drive approaches are ready. However, it is likely that most of this delay will be because of the need to adequately engage with stakeholders, and time required to effectively address regulatory requirements, and only minimally due to technology development reasons (Glover et al., 2018).

4.5 NEED FOR REGIONAL COLLABORATION

Border issues constitute one of the greatest impediments to African Science, Technology and Innovation (STI) and economic development. These boundaries limit the movement of researchers, materials and equipment across borders. They discourage collaboration and promote costly and ineffective duplication of effort. In Africa, there have been some excellent examples of cross border collaborations, which are proving quite productive and effective, in particular, the efforts of the NEPAD Agency's African Biosciences Initiative (ABI) networks. Without broader policy and programme harmonization, the hope for countries in moving towards knowledge-based economies will not be feasible (NEPAD, 2018).

Since transgenic mosquitoes are organisms that are mobile and having the transgenic construct will enable spread through interbreeding populations, which cannot be contained within a single country, transboundary issues therefore have to be taken into consideration in the regulation and deployment of transgenic insects. In this case, a regional regulatory approach will be relevant instead of only having independent national regulatory systems as has been the case with GM crops.

Africa in recent years is building strong collaborations among countries and harmonizing technical requirements and processes for regulating transgenic insects, the most practical being at the regional level. These regional platforms have exhibited many advantages. For example, resolutions are closer to being achieved as a result of sufficient motivation amongst the concerned countries.

Angola, Botswana, Mozambique, Namibia, South Africa, Swaziland, Zambia and Zimbabwe have formed a partnership, Elimination 8, to collaborate across borders for the attainment of a bold goal by 2030. These eight countries are leading the way for the African continent, demonstrating that elimination of malaria is possible in continental sub-Saharan Africa. The Technical Committee deliberates on the technical and strategic elements of collaboration, guiding policy standards, and bringing new evidence and operational approaches to improve national and regional elimination strategies (Southern African Development Corporation, 2016). African regulators can indeed take the lead in transgenic mosquito regulation as guidelines for implementation of transgenic mosquito/gene drive are being drawn and harmonized across African regional networks.

4.6 RISK ANALYSIS

In the course of regulating transgenic mosquitoes and gene-drive technologies it will be essential that risk assessments are undertaken at different stages to: identify any potential risks that may arise in relation to the technology; allay any stakeholder fears in relation to potential negative impacts of the technology; and identify specific mitigation measures to address any negative impacts that may arise during development or implementation of the technology. Risk assessments will always be conducted from the perspective of the accrued benefits versus potential risks, ensuring to not overemphasize one over the other.

Expert views currently indicate that no major risks are foreseen that cannot be mitigated in the application of gene-drive technology for malaria elimination, and that the potential benefits will almost certainly outweigh any minor risks observed (Ricroch, 2019). However, regulators are vigilant and would update available information on the state of development, risks and benefits of the technology.

4.7 CONCLUSION

Regulation of gene-drive technology is unique because it would translate from transgenic insects for vector control directly introducing mortality through the elimination of malaria in Africa. The regulatory processes in Africa are considering multiple aspects, ranging from technologies initiated by agricultural genetic modification processes, through transgenic insect vector control, to health value propositions. This will require bridging from agriculture to health and the regulatory pathway being targeted will form a breakthrough model for other emerging technologies that need to be harnessed for economic development in future (Glover et al., 2018).

The New Partnership for African Development (NEPAD) Agency, through the African Biosafety Network of Expertise (ABNE) and the African Medicines Regulatory Harmonization (AMRH) and by virtue of its position in the African Union, Regional Economic Communities (REC) and national systems, play an enabling role by coordinating processes and programmes at all these levels. The New Partnership for African Development Agency supports policy implementation and strengthening of regulatory systems, facilitates the preparation of appropriate guidelines and supports the fora for experts and regulators engagements at both national and regional levels (NEPAD, 2018).

The release of genetically modified mosquitoes will present several challenges, which may not necessarily be unique but probably not pursued with other vector control interventions, including biological control agents without genetic modification (Friedman et al., 2020), likelihood of mosquitoes crossing national boundaries, for instance will require information sharing between countries. Besides, fears of disturbing biodiversity will require risk assessments and mitigation plans. In addition to assessments conducted by the developers and regulators, independent risk assessments may be essential to improve perception of non-bias among stakeholders and should be conducted at the different stages of development. Data from these assessments should then be made publicly available to improve transparency and compliance from all parties involved. Moreover, regulatory systems across the continent, including those already available for regulations of genetically modified crops, may need to be adjusted to effectively regulate gene-drive mosquitoes. Nonetheless, several efforts are underway to ensure conditions are created that would enable the deployment of the technology as soon as it is eventually ready.

REFERENCES

Bhatt, S., Weiss, D., Cameron, E., Bisanzio, D., Mappin, B., Dalrymple U., and Eckhoff, P. (2015). The effect of malaria control on *Plasmodium falciparum* in Africa between 2000 and 2015. *Nature*, ***526***(7572), 207–211.

Bongomin, O., Ocen, G. G., Nganyi, E. O., Musinguzi, A., and Omara, T. (2020). Exponential disruptive technologies and the required skills of industry. *Journal of Engineering*, ***2020*** Article ID 4280156, 17 pp.

Durnez, L., and Coosemans, M. (2013). Residual transmission of malaria: an old issue for new approaches. *Anopheles Mosquitoes—New Insights into Malaria Vectors*, 21, 671–704.

Eckhoff, P. A., Wenger, E. A., Godfray, H. C. J., and Burt, A. (2016). Impact of mosquito gene drive on malaria elimination in a computational model with explicit spatial and temporal dynamics. *Proceedings of the National Academy of Sciences*, ***2016***, 11064.

Elliott, R. (1972). The influence of vector behavior on malaria transmission. *The American Journal of Tropical Medicine and Hygiene*, ***21***(5 Suppl), 755–763.

Friedman, R. M., Morris, J. M. and Akbari, O. S. (2020). Gene drives: new and improved. *Issues*, ***36***(2), 1.

Gantz, V. M., Jasinskiene, N., Tatarenkova, O., Fazekas, A., Macias, V. M., Bier, E., and James, A. A. (2015). Highly efficient Cas9-mediated gene drive for population modification of the malaria vector mosquito Anopheles stephensi. *Proceedings of the National Academy of Sciences*, ***112***(49), E6736–E6743.

Glover, B., Akinbo, O., Savadogo, M., Timpo, S., Lemgo, G., Sinebo, W., Akile, S., Obukosia, S., Ouedraogo, J., Ndomondo-Sigonda, M., Koch, M., Makinde, D., and Ambali A. (2018).Strengthening regulatory capacity for gene drives in Africa: leveraging NEPAD's experience in establishing regulatory systems for medicines and GM crops in Africa. *BMC Proceedings*, ***12***, 11.

Hammond, A., Galizi, R., Kyrou, K., Simoni, A., Siniscalchi, C., Katsanos, D., and Russell, S. (2016). A CRISPR-Cas9 gene drive system targeting female reproduction in the malaria mosquito vector *Anopheles gambiae. Nature Biotechnology*, ***34***(1), 78.

Jobin, W. (2014). Suppression of malaria transmission and increase in economic productivity in African countries from 2007 to 2011. *Malaria World Journal*, ***5***(4), 1–8.

Marshall, J. M., and Taylor, C. E. (2009). Malaria control with transgenic mosquitoes. *PLoS Medicine*, ***6***(2), e1000020.

Morris, E. J. (2014). *Biosafety Regulatory Systems in Africa. Biosafety in Africa: Experiences and Best Practices*, Michigan State University Press, East Lansing, MI, 7, 64–77.

NBMA (2015). *National Biosafety Management Agency Act*. https://nbma.gov.ng/wp-content/uploads/2021/04/National_Biosafety_Management_Agency__Act_2015-signed.pdf

NBMA (2019). *National Biosafety Management Agency Act, as Amended*. https://nbma.gov.ng/wp-content/uploads/2021/04/Amendment-of-NBMA-Act-2019.pdf

NEPAD (2018). *Gene Drives for Malaria Control and Elimination in Nigeria. High Level African Panel of Emerging Technology Report*. https://www.nepad.org/publication/gene-drives-malaria-control-and-elimination-africa

Pates, H., and Curtis, C. (2005). Mosquito behavior and vector control. *Annual Review of Entomology*, ***50***, 53–70.

Ricroch, A. (2019). Global developments of genome editing in agriculture. *Transgenic Research*, ***28***, 45–52.

Southern African Development Corporation (2016). *Annual Report 2016 of the Elimination 8 Partnership: Working towards a Malaria Free Southern Africa*. SADC, Gaborone, Botswana.

WHO (2017). *World Malaria Report 2017*. World Health Organisation, Geneva.

WHO Vector Control Advisory Group. (2017). Fifth meeting of the vector control advisory group, Geneva, 2–4 November 2016. World Health Organization, WHO/HTM/NTD/VEM/2017.02(Licence CC BY-NC-SA-3.0 IGO).

5 Ethical and Social Aspects of Modern Biotechnology

Charles Oluwaseun Adetunji, Kingsley Eghonghon Ukhurebor, and Olugbemi Tope Olaniyan
Edo State University Uzairue

Olugbenga Samuel Michael
Bowen University

Julius Kola Oloke
Ladoke Akintola University of Technology

Benjamin Ewa Ubi
Ebonyi State University
Tottori University

CONTENTS

5.1 INTRODUCTION

Bioethics conventionally emphasizes on creating moral bounds amongst various kinds of acts in the biological and medical fields (Martensen, 2001; Bracanovic, 2012). The word "bioethics" is usually seen from two different perspectives originating in the 1970/1971 from two different scientists, Van Rensselaer Potter and André E. Hellegers:

DOI: 10.1201/9781003179177-5

> Van Rensselaer Potter (1971) advocates an inclusive and universal understanding of bioethics that assimilates even environmental ethics in a broader context (Potter, 1971); and André E. Hellegers applied the word "bioethics" in the academic, public policy and the human biological sciences context for the first time.
>
> *(Reich, 1995)*

According to Hellegers, bioethics is a means of determining rights and avoiding impending ethical conflicts caused by an innovative perception of biological and medical sciences; this supplementary limited interpretation has turned out to be leading in several philosophies and in the practice of bioethics (Reich, 1999; Bobyrov et al., 2012). Presently, bioethics has been recognized as a field of study that offers a critical, practical perception on medical science and modern biotechnology, as well as the conventional or traditional context of human rights. Consequently, authors like Brooke (2008) were of the opinion that bioethics should be categorized together with queer and critical race theory as well as cultural studies and multiculturalism as a significant method to the general human rights (Muldoon and King, 1995; Karsjens and Johnson, 2003; Brooke, 2008; Hoberman, 2016; Saxén, 2017).

Indisputably, bioethics outlines the delineations of fundamental rights twofold:

- It extends the catalogue of fundamental privileges or at best proposes to spring the understanding of privileges to the field of bioethics.
- Bioethics programmes spread to the topics of safety and protection. For instance, this is applied to forthcoming generations and implies that bioethics covers both medical and biological sciences as well as legal, logical, ethical and philosophical backgrounds of the individual.

The perception of bioethics progressively serves as an instrument for framing and understanding several evolving biomedical machineries and assists in the assessment of their ethical and legal consequences. Bioethics as a field of study is comparatively novel and so its hypothetical situations, legal understandings and strategy implications are hardly recognized among the judicial aspects or adequately particularized to be broadly used in judicial procedures. Hence, law courts habitually constrain their situations to bioethics dissertation to a simple citation of several binding and non-binding gadgets, even ignoring their opportunity and context. Nevertheless, if we carefully examine the dissertation on the jurisdictional subjects of rights in biomedical sciences, a substantial extension of the aspect could be detected. Reverence for self-respect as well as the privilege to confidentiality is employed in the interpretation of the choices on human biological constituents, nucleic acids (specifically DNA) samples, choices over the protective privileges of in vitro embryo, as well as in influencing the privilege to choose what kind of studies are to be carried out on biological samples which were previously collected. The judiciary sector globally is confronted with these intricate issues about the evolving bio-social individualities of humans; and in this aspect, bioethics appears to stimulate substantial preparatory efforts via the exploration of the basic ethical issues and consequences of innovative biotechnologies. However, some of these apprehensions are known as fundamental privileges, but bioethics assists as a vital test site in the crystallization of the fundamental norms and procedures for understanding innovative machineries in the field of biological sciences (Karsjens and Johnson, 2003; Bagheri, 2014; Hoberman, 2016; Saxén, 2017).

Supposedly, law ought to avoid two extreme situations in regard to these innovative technologies: firstly, it ought to side-step the over-guideline and untimely guideline, which occurs when rule jumps too swiftly to the novel systematic developments deprived of adequate period for replication on the ethical as well as the social consequences of an innovative technology; and secondly, frequent critical situation has its foundation in some legitimate backgrounds in strongly separating the idea whereby the adjudicators refrain from going into "science" (Karsjens and Johnson, 2003; Hoberman, 2016; Saxén, 2017; Builders et al., 2019).

Presently, bioethics as a field is addressing a very broad aspect of human investigation, such as the scientific arguments about the boundaries and restrictions of life (like euthanasia and abortion), surrogacy, the distribution of limited medical resources (like medical rationing and donation of organs) and the belief to refuse medical attention due to traditional, cultural or religious motives.

Nevertheless, those involved in bioethics sometimes disagree within themselves due to the specific bounds of their field or discipline, arguing if the aspect should concern itself with the ethical assessment of all enquiries concerning biological and medical sciences, or just a subsection of these enquiries (Bracanovic, 2012; Horne, 2016; Hoberman, 2016), while some would extend the possibility of ethical assessment to embrace the principles of all dealings that could assist or cause impairments to living organisms proficient of sensation anxiety.

It has been posited that the scope and boundaries of bioethics could increase with increasing advancements in modern biotechnology including cloning, gene therapy, extension of lives, genetic engineering in humans, astronomical studies and astroethics (Freemont and Kitney, 2012). Agreeably, these technological advancements would influence future developments and evolution, and may need innovative ideologies underlying the fundamental aspects of life, like "biotic ethics" that respects life at its fundamental biological procedures and structures (Mautner, 2009; Kuiken, 2011; Horne, 2016; Hoberman, 2016; Mishra et al., 2020; Adetunji and Anani, 2021a, 2021b; Adetunji et al., 2021; Anani and Adetunji, 2021; Anani et al., 2020; Dwivedi et al., 2021; Islam et al., 2021; Olaniyan and Adetunji, 2021; Thangadurai et al., 2021).

5.1.1 Basic Principles and Boundaries of Bioethics

Those involved in the field of bioethics preferably use four main principles when assessing the advantages and complications of/in the medical and biological actions. Beauchamp and Childress (2013) state that the common morality or ethics "refers to norms about right and wrong when dealing with human conduct which are broadly shared because they become a stable social compact". Apparently, for biological and medical exercise in order to be considered ethical, it should reverence all these four principles of medical ethics as postulated by Beauchamp and Childress "Autonomy, Justice, Beneficence and Non-maleficence" (Entwistle et al., 2010; Horne, 2016; Friesen et al., 2017). However, the use of technology-aided reproduction raises inquiries in each of these aspects.

5.1.1.1 Autonomy

This aspect entails that there is self-sufficiency of thought, purpose and dealings when making decisions concerning medical and biological dealings especially on the side of the clients. Consequently, the administrative procedure should be free of intimidation or persuasion. For clients to make copious informed decisions, they should know all hazards and benefits of the process as well as the prospect of achievement.

5.1.1.2 Justice

The impression that the problems and the assistance of innovative or tentative dealings should be evenly disseminated among all categories of persons. This aspect entails that the measures sustain the essence of present rules and are realistic to all involved. The medical and biological scientists should reflect on these four core aspects when assessing justice: "fair distribution of scarce resources, competing needs, rights and obligations, and potential conflicts with established legislation". Technology-aided reproduction generates moral problems mainly because dealings are not consistently obtainable by all persons.

5.1.1.3 Beneficence

This aspect entails that the process should be provided with the determination of undertaking decency for the clients. The demands that medical and biological scientists advance and preserve aids and knowledge, incessantly apprise knowledge via training, reflect the individual situations of all clients and attempt for net advantage.

5.1.1.4 Non-Maleficence

This aspect entails that the process hardly causes impairment to the clients involved or even other persons in the society. For example, infertility experts work under certain assumptions that they are undertaking no impairment or at best curtailing impairment by trailing the larger benefits. However, since assistive reproductive machineries have restricted achievement rates with ambiguous general consequences, the emotional state of the clients could be undesirably obstructed. In several instances, it is hard for medical experts to effectively employ the impairment principle.

Other basic principles which have now been incorporated to this list of basic principles are "non-maleficence, human dignity and the sanctity of life" (Friesen et al., 2017). The "Belmont report of 1979" has assisted studies in a route that focuses on the protection of susceptible aspects and assisted in the fight for transparency amongst researchers and the subject in question (Friesen et al., 2017). Agreeably, there is a significant growth in research presently as a result of technological advancement (Friesen et al., 2017). Hence, there is a need to revising these lists of basic principles to suite the present situation. Consequently, the "Bioethics Society of Ohio State (2013)" reported that there are other important principles of bioethics such as "placement of value on discussion and presentation"; as such, these other principles also need to be given appropriate attention.

5.2 EXISTENCE OF BIOETHICS IN BIOTECHNOLOGY

5.2.1 Food and Agriculture

Food has been identified as one of the major requirements for the sustainability of mankind and right of entry to it are essential ethical questions. Great access to good food has been identified as a human right but there is a high level of hunger that affects mankind which has led to increase of mortality and suffering. According to the United Nations (1948), food accessibility is a conception encompassing from "Article 25 of the Universal Declaration of Human Rights which declares, that everybody possesses a great privilege to a standard of living sufficient for the well-being and health of himself and of his family, including food...". Despite all these, there are over ~800 million persons suffering from higher levels of malnutrition and hunger worldwide. This is an unacceptable condition, and heroes of biotechnology have faith that the application of biotechnology could help in the mitigation of all the aforementioned challenges, and most especially in the area of food insecurity (Serageldin, 1999).

Biotechnological approaches have gained attention because of its utilization of living entities or components within organisms to produce or adjust a product, to improve plants or animals, or modify microorganisms for specific purposes, and cautious evaluation is needed to ascertain the risks involved with the potential formation of new selection initiated by the use of biotechnological tools like genomics, bioinformatics, molecular breeding, transformation, diagnostics, and vaccine technology. Acceptability of genetically developed foods has limited agricultural biotechnology advancement. Ignorance on biotechnology has amplified uncertainties in the minds of end users on the merits, doubts, health risks and detrimental effects, and ecological concerns (Sawicka et al., 2020).

Food safety was defined during the "World Food Summit 1996 Rome declaration on food security", where it was stated that food security occurs when the majority of people at all times have economic access to adequate, safe and nutritive food which could help in meeting the nutritional requirement (FAO, 1996). There is a need to evaluate the reason for the high level of food insecurity. Food insecurity occurs as a result of the high level of poverty, inadequate levels of human and material resources, and additionally the incapability to suitably use the existing human and material resources, deficiency of evidence and training, etc. (Wight et al., 2014).

Biotechnology is viewed as a sustainable technology that could help in providing the necessary solution to all the aforementioned challenges of food insecurity. This will go a long way towards providing adequate food for the ever-increasing global population, and also mitigate against all the erroneous impression that "Green Revolution" could not meet all the agreements in the long term. Also, post "Green Revolution" apprehensions on yield slowing down, yield ceiling, decrease in the level of production, large yield gaps, unsatisfactory dissemination of food and properties and environmental dilapidation as explanation for their absolute denunciation. The second line of thinking is that innovative biotechnology gives prospect for different types of species most especially in the environment. It has been identified as an essential tool for different types of climatic and land conditions with low or high efficiency;

targeting it to come across any environmental situations would be more reasonable. However, these are established on unscrutinized norms. It has been discovered that some benefits that could not be provided by the Green Revolution has been properly taken care of through the new biotechnology such as production of high-yielding staple crops, herbicide-resistant crops, increase in food quality, insect-resistant crops, salt tolerant crops, virus-resistant crops, etc. (Azad and Hossan, 2004; Pingali, 2012).

Ethically, there is a high tendency to state that biotechnology portends the capability to increase the level of food quality and increase in it supply and production without having any adverse effect on the ecosystem as well as the numerous agroecosystems, which possess the tendency to feed most of the hungry people round the globe. All these recent developments have nullified all the false assumptions about the practical application of biotechnology as a sustainable technique that could lead to increase in food production. The application of terminator technology has been applied for effective development of "FLAVR SAVR tomatoes which are genetically modified" in order to meet numerous market requirement. Moreover, another controversial issue involves the application of Genetic Use Restriction Technologies (terminator seed technology) for small-scale farmers most especially in the non-industrialized countries that need to buy recent seeds almost every year. One of the great issues here is that farmers need to save and utilize seeds obtained from preceding plants for the forthcoming year. Nevertheless, it is doubtful on the cost efficiency of utilizing saved seeds, that have a greater tendency to become contaminated with disease as a result of elongated and inappropriate storage or grow disease far along in the sequence, as well as show significant drop in hybrid vigour (Kramer and Redenbaugh, 1994; Mobasser et al., 2012).

In several industrialized nations, large-scale agriculture is performed and agricultural activities have been identified as a very prosperous business where agriculture is recognized as a social way of life because it is mostly performed at a smaller scale. One of the greatest problems is how to merge the situations in the developed and underdeveloped countries. Globalization has been discovered to have its merits and demerits. Enhancement in the level of market competition and inadequate resources to establish improved infrastructure most especially for agricultural dealings has threatened the prospect of small-scale agriculturalists in underprivileged regions. However, it has been argued that access to adequate information on the application of relevant technology has become comparatively more straightforward with globalization (Anderson, 2010; Rowe et al., 2012).

Also, many members of the public have negative believe on genetically modified foods with several ethical issues including the need for labelling. Numerous surveys have established the fact that certain members of the public have bad feelings against genetically modified foods, most especially in Japan and Europe. Most of the poor people do not base their choice on the nature of the food whether it is genetically modified or non-genetically modified. It has been validated that genetically modified food portends the capacity to increased nutritional content of food or when they are consumed as edible vaccines that could be applied for alleviation nutrient-deficient diseases. These have a greater application in the countries, most especially where they have a larger population suffering from malnutrition (Maghari and Ardekani, 2011; Dizon et al., 2016).

5.2.2 Health and Medicine

The perception of medicine and health has a lot of relationship with the environment and food. It has been observed that malnutrition and hunger could lead to a high level of mortality and sickness. It has been established that the application of biotechnology has diverse application especially in the medical sector which could serve as another major issue since the invention of sequencing of genomes which could be partly linked to numerous disagreements over the Human Genome Diversity Project. Numerous people have asked several questions on the potential of biotechnology as a sustainable tool towards the provision of adequate health care to most sick people. Some other people believe that it might increases the gap between the rich and the poor people (Silverman, 2004; McGuire et al., 2008; Welcome, 2011).

The application of molecular medicine might be very costly towards the achievement of adequate health care which might hardly be the first option in the treatment of patients most especially in developing countries where millions of people lack adequate access to basic health care. The issue of fierce patent protection and poverty have increased the level of the unavailability of the necessary drugs most especially to people in the developing countries. The issue of high level of antibiotic resistance in microorganism have increased the needs to search for a new drug most especially in the developing countries which make most of the people that dwell there to require new drugs for the management of these diseases that have developed high resistance to the available antibiotics (Blackstone et al., 2014; Ayukekbong et al., 2017).

The application of traditional medicine has been realized as a sustainable technique for the management of numerous diseases most especially the development of Acupuncture and Ayurveda in India and China (Rastogi and Kaphle, 2011; Fokunang et al., 2011). Most people in these countries use this medical technique as a great source of income, while it is also a common cultural practice in these regions. Numerous developing countries have developed more cost-effective generic drugs with have been identified in having very huge market in poor nations, most especially in Africa. Moreover, Mackenzie (2001) reported that in June 2001, the UN made a pronouncement that a particular Global Health Fund of $10 billion was released for the management of malaria, AIDS and tuberculosis diseases that are prevalent in most of these poor countries.

Furthermore, it was observed that most of these arguments were not made to denigrate patents. Patents could inspire technical feelings and maintain eagerness of investors to reconnoitre innovative ideas. One of the greatest problems involves how we can strike a balance on the great necessity of the poor with satiating the request of the manufacturers of drugs. Numerous scientists have invested several amounts of money, time, intelligence, and the discovery that has been patented is their greater reward. Nevertheless, Bhardwaj (1999) reported that there are restrictions to self-sufficiency and technical independence. Researchers also have collective responsibility above their individual independence to deliver from their proficiency without undertaking impairment in any form.

Therefore, there is a need to strike a balance between people that dwell in the less-privileged countries and advanced nations so that people in the underdeveloped countries could benefit from all these scientific advancements. There should be provision of adequate health-care system that is associated with biocentric intelligence when animals are utilized for investigation, which is an alternative reason of apprehension in numerous nations. Disparagement has been upturned in contradiction of international drug companies utilizing poor, illness predominant nations as testing grounds or for palliative control trials, particularly in African region where it is very costly to perform such trials; also, since the wide-ranging is at critical frightening phase, the results for the trials are stress-free to obtain. Ethically, we must deliver medicine and food to most people that require them. The perception can be contended from down-to-earth ethics, which articulates better for maximum individuals or established on philosophy of "Justice by Rawls", that discusses for goodness established on the minimum that may well be profited.

5.3 RECENT DEVELOPMENTS IN BIOETHICS

Kirsten Martin et al. (2019) described how no attention is given to the dissemination, development, and design of technology in most firms. Recently, increase in the growth of technological developments has been witnessed; hence, there should be need to identify firms that owe the society. The authors focused attention on digital technologies and how firms should implement the use of bioethics in their deployment and development. Ugbogbo and Atu (2016) revealed that technological advancement such as networking, information distribution and cybernation will also have corresponding challenges like bioethics, political and social issues. The authors utilized the library-based approach to evaluate bioethical issues of information technology (IT). It was discovered that among other ethical issues identified, health and biological concerns were raised as possible challenges of IT. Suggestions were made in terms of putting up legal frameworks to address some of the issues. Strategies and guidelines should be applied so that general information can be accessed in an ethical manner.

Studies have shown two different opinions on the bioethical issues in science and technology. On one side, it is believed that science and technology are bio-ethically neutral except its knowledge used for wrong means (Iaccarino, 2001). The other side viewed science and technology in terms of generation of knowledge as not free of ethical values and norms (Iaccarino, 2001). The authors revealed that one of the controversial areas of science and technology concerning bioethical issues in the 20th century is called "techno-scientific systems" where knowledge is generated and applied for example in the development of atomic bomb or biological weapon (Marris et al., 2014; Jefferson et al., 2014). Also, the problem of human beings, animals as experimental subjects are critically analysed with emphasis on the impact of technology, science, plus techno-science on the environment and the society.

Alyeksyenko et al. (2019) described the impact of legal regulation and safety of biotechnology as an emerging branch of science and technology concern with areas of transfer and utilization, development and implementation. The authors investigated the mechanism to implement bioethics policy according to international standards to assist as a template for the development of biomedical laws via incorporation of bioethics moralities. Their aim is to view and analyse the extent and practice of bioethics in line with international regulations and laws. From their findings it was discovered that standards and regulations are present in current international arrangements on the issue of biotechnology development from bioethics point of view. Hence, it was noted that these aspects must form the foundation of the Code of Bioethics and international biomedical law.

Studies have shown the importance of bioethical values and norms during the regulation of biotechnology because the principles of bioethics are generalized across the globe, but the practice are not generalized due to varying principles and belief systems. Hence, it cannot be seen as a law at the international level. Therefore, in unifying biotechnology practice, there must be adoption of the legal framework on bioethics. Over the years, various laws have been formulated such as

- The first laws were: "Nuremberg Code (August 1947, Nuremberg)".
- The "Helsinki Declaration of the World Medical Association Ethical Principles of Medical Research with Human Subject", approved in 1964 at the 18th Assembly.
- The "International Ethical Guidelines for Biomedical Research on Human Beings", approved in 1982 by the "Council of International Scientific and Medical organizations" (amended in 1993 and 2002), and other laws.
- The "Convention on Human Rights and Biomedicine (1996)".
- The "Universal Declaration on the Human Genome and Human Rights (1997)".

Presently, studies have shown that only Nuremberg Code is enforced (Shuster, 1997). It should be noted that the control of biotechnology is influenced by the above outline legal act at international level. Some treaties also consolidated the bioethical issues in utilizing biotechnology for research such as "the Convention on the Prevention and Punishment of the Crime of Genocide (1948), the UN Convention on Human Rights (1989), plus the International Covenant on Human Rights (1966)". Several bioethical moralities were articulated in Tokyo at "the IV International Congress on Bioethics" concerning biotechnology such as the human right to self-resolution of matters involving psyche, recognition of individual's autonomy, right to free informed consent, emotional status, fulfilling the Hippocratic oath and its extension, the principle of justice as equal admittance to social aids. Currently, there is a universal declaration on preventing unethical use of cloning in humans for reproductive purposes which also extend to every individual conducting research in biomedical science like "the Human Genome and Human Rights, Convention on Human Rights and Biomedicine (1997), UN Declaration on Human Cloning (2005)". It is mandated that all human biomedical study ought to be subjected to an independent moral review ethics Committee to evaluate all the techniques before being practised. The "World Conference on Women (Beijing, 1995), UN Cairo Convention on Democracy and Development in 1994" enshrined some reproductive rights like infertility treatment, right for decision on reproducing offspring, right on reproductive health. Generally, in assisted reproduction and reproductive biotechnology, there is no legal act, law or regulations at the international level, even with surrogate maternity. Formulating a single international law concerning reproductive biotechnology in assisted technologies will help to regulate and resolve controversies. Also, Cartagena Protocol act was created to regulate all issues of genetically modified organism's transportation from one state to another which are seen as common practice in food and pharmaceutical industries and agriculture. Hence, there must be regulation as more genetically modified products are being created and utilized without understanding the long-term biological effects and environmental consequences.

Rui (2014) revealed that bioethics in the medical field has huge influence on the current developments and evolution. The fundamental human right and dignity should be respected such as privacy, self-determination and integrity making the patient the centre of attention. Okan et al. (2012) revealed attention today has been focused on individual-oriented political, socio-economic and intelligent means despite the use of science and technology to solve human problems. Environmental education over the years has tried to prove that man is not the only element of nature; therefore, in living harmoniously with other elements, there must be a regulation in human activity, which makes bioethics a powerful tool in implementing and regulating relevant processes in science and technology. Benatar et al. (2003) also revealed in their study that most of the issues in science and technology today became topics of controversies as a result of advancement in biotechnology and proper bioethical framework and legal regulation must be adopted to handle those issues. Motoko et al. (2014) suggested that serious caution must be taken when utilizing genome editing technologies in biotechnology. Several bioethical concerns are now being raised for discussion and this has resulted to the report issued by "the National Commission for the Protection of Human Subjects of Biomedical and Behavioural Research (Belmont Report) and United Nations Universal Declaration on Bioethics and Human Rights concerned with research on human subjects in 2005 dealing with justice, consent, confidentiality and privacy". Others are social responsibility, non-discrimination protection of biosphere, biodiversity, future generations and the environment. European countries still frown against the cultivation of genetically modified crops with all the attendant benefits. Advancement in biotechnology has become a serious concern even in United States where court rulings are implemented on many genetically manipulated products like gene therapy, transgenic plants or animals, cloning plus recombinant DNA technology. The National Institutes of Health produced a guideline on recombinant DNA technology to regulate the utilization against chemical and germ warfare, antiviral and antibiotic drug resistance, and use. Today, DNA can be obtained from an extinct species, produce an embryo by nuclear transplantation and inseminated into a female surrogate animal for delivery. Human germline through pre-implantation genetic screening can be altered if there are certain genetic disorders like phenylketonuria under the parent's consent. Many bioethics issues have also been raised concerning stem cell study predominantly the use of embryonic stem cells in regenerative medicine where an embryo must be destroyed to obtain a stem cell.

Sarah et al. (2015) revealed that synthetic biology which is an aspect of biotechnology such as synthetic medications and muscle tissue for biofuels is becoming popular today for sustainable economic value and bioethical concerns are also generated. The authors evaluated the bioethical issues linked with synthetic biology and their relevance globally in line with the environment and human health. The fundamental issues linked with bioethics are prospects, present rules and regulations, future advancement, morality, legal framework, religious aspects, safety and research. Cascais (1997) suggested that comprehensive review of emerging issues warranting bioethics should inform us about the difficult nature that envelops its development. The author highlighted few emerging issues such as availability of new biomedical technologies, abuses in experimentation on human subjects, purpose and meaning of medical care, genetic engineering, environmental health, reproductive physiology, behavioural manipulation, thus bioethics is best defined based on practical and theoretical inquiries not just by individual opinion.

Subrata (2004) revealed that biomedical engineering advancement and progress in medical science has helped develop a lot of novel medical technology of bioethical concerns. The authors highlighted issues like funding, clinical trials of new implants, drugs or devices, human or animal experimentation and conflicts of interest. Therefore, biomedical engineers must uphold the highest standards of professionalism in their practice and as technological advancement move to the next generation, every biomedical scientist must sort out the bioethical issues in bringing the technology to the table. Pamela Andanda (2009) showed that in Africa where agricultural sector contributes to the major economic value chain, biotechnology in relation to pharmaceuticals and industrial sector is still in their early stage of development, but agricultural sector has shown tremendous progress; therefore, effort is being put in place for its regulation and adoption. Andanda (2006) evaluated the efforts made so far to develop a legal framework for genetically modified organisms by Kenya and South Africa. They discovered that a holistic approach should be adopted to handle many controversies in genetically modified organisms in order to gain public trust and maintain bioethical practice.

5.4 DEALING WITH ETHICAL ISSUES ARISING FROM INTEGRATED TECHNOLOGIES

Ethical issues around biotechnological developments will remain a source of intense argument between the scientific community, government, manufacturers and the general public. Ethical concerns or reflections have helped shape the guidelines for biotechnological advancement such that it can be accepted in the society due to deep consideration for human dignity, ecosystem interference, etc. (Rosie, 2004). Genetic manipulation in plants and animals as a result of biotechnology are major sources of ethical argument for some decades now in the medical and agricultural practice. Establishment of the Biotechnology and Human Genetics commissions, two bodies saddled with such responsibilities of looking into ethical issues around biotechnological advancement by the UK government, represents the willingness and involvement of the government in the regulation and policymaking for biotechnology. Interestingly, the views, opinions, interpretations and translations of the ethical concerns caused by biotechnological advancement are varying across diverse societies (Polkinghorne, 2000).

The gene manipulations involved in the enhancement of crop breeding using methods like embryo rescue and mutagenesis; transfer of genes from plants to animals and vice versa irrespective of evolutionary distance using recombinant DNA technology has been seen by many as a distortion of the fabric of life, while scientists see it as means of achieving genetic diversity for such organisms with the desired or durable agricultural characteristics which might been impossible without such biotechnological advancements. Since biotechnological advancement have both merits and detrimental effects, it is not out of place to be concerned about the risks posed by the technology of moving genetic materials across diverse species (Prince Charles HRH, 2000).

Controversial issues in industrial biotechnology advancement can be rightly exemplified by the production of vanillin, algae-based oil and artemisinin. The case of semi-synthetic artemisinin (SSA) production was criticized by Canadian-based ETC Group that it will put farmers who make natural artemisinin through the production of wormwood out of job thereby taking away their means of livelihoods (Heemskerk et al., 2006; Peplow, 2016). Furthermore is the synthetic production of vanillin from yeast through metabolic reprogramming. Although the vanillin produced through this process was chemically close to the natural vanillin but once again the Canadian-based ETC Group together with Friends of the Earth USA criticized this technological advancement that the synthetic vanillin is not natural and as such not sustainable due to the involvement of genetic engineering in its production. Also, they claimed that the livelihoods of agriculturalists who harvest in a natural and sustainable way with regard to the ecosystem will be jeopardized (ETC Group, 2013; Bomgartner, 2016).

Interestingly these disputes were due to opinions and social value systems which are centred around five major concerns in the society pertinent to the advancement of biotechnology and they include naturalness, sustainability, risk management, economic justice and innovation trajectories.

5.4.1 The Case of Vanillin Production

It has been documented that a chemically fabricated vanillin generated through yeast whose metabolic pathways has been in comparison to the Amyris platform by the Swiss-based company Evolva in 2011. The most bulging component available in the vanilla is vanillin. Furthermore, most of the vanillin present in the market is generated through petrochemical processes. Also, it has been observed that the vanillin derived from the synthetic approaches is more cost effective when compared to the one obtained from the natural vanilla sourced from the vanilla orchid, that constitutes up less than 1% of the vanillin that are applied in some of the recent production (Bomgartner, 2016). The company Evolva believed that the vanillin generated from the application of yeast which are natural could be sustainable and could give better quality when compared to the other synthetic vanillin. The level of sustainability could be guaranteed because it does not depend on paper pulp or fossil resources like other artificial vanillin. The love of the quality of the vanillin was very high because it has a very high level of the same chemical structure available in natural vanilla. The vanilla is produced through fermentation process which makes the process natural which is referred to as natural food production process under the US or EU law. Conversely, other environmental organizations such as Friends of the Earth USA together with the ETC Group had a conflicting agreement to this related claims and innovation (ETC Group, 2013). They are not against the quality of the product, but they are after the product which is not natural. Its production might not be sustainable due to the fact that they might have been fabricated through the application of highly engineered organisms. Moreover, they had a belief that the production will destabilize the incomes of vanilla farmers who could not guarantee the sustainability of their product most especially when linked to their direct natural environment. It has been observed that the two environmental organizations make an assertion that biotechnological processes do not have any respect for local ecology because they could change any crop into a feedstock most especially for their processing facilities while the adoption of the 'natural' value chain need to be in agreement with the indigenous ecology and supports it conservation in order to allow for its continual production. Therefore, agreeing to the environmental organizations, this old-fashioned value chain is improved for the protection of friable rain forests.

5.5 CONCLUSION AND FUTURE PERSPECTIVES

This chapter has provided detailed information on the numerous techniques, social, associated ethics, economic factors that could improve and evaluate the merits and associated risks before the application of the modern methods of biotechnology. Biotechnology has also played a crucial role in the provision several beneficial cellular and molecular biological process with wide applications in animal and plant improvements for sustainable food and agricultural production, as well as high effectiveness for discovery of novel drugs in biomedical sector and value-added products. It was also discovered that ethical matters in the biotechnological fields are hardly restricted to the matters that are mentioned in this chapter. Therefore, there is need to evaluate judgements on some of the issues regarding the usefulness of biotechniques for sustainable advancements in various aspects of human life. We need to establish adequate balance with the society on required expertise and knowledge and the role they play in the biotechnological inventions. It is important to consider bioethical principles in relation to modern biotechnology on several issues such as how to accredit value of human well-being, health, food, maintaining ecosystem balance, as well as other socio-economic considerations.

REFERENCES

Adetunji, C. O. and Anani, O. A. (2021a). Recent advances in the application of genetically engineered microorganisms for microbial rejuvenation of contaminated environment. In: Adetunji, C. O., Panpatte, D. G. and Jhala, Y. K. (eds), Microbial Rejuvenation of Polluted Environment: Microorganisms for Sustainability, vol 27. Singapore: Springer. https://doi.org/10.1007/978-981-15-7459-7_14

Adetunji, C. O. and Anani, O. A. (2021b). Utilization of microbial biofilm for the biotransformation and bioremediation of heavily polluted environment. In: Panpatte, D. G. and Jhala, Y. K. (eds), Microbial Rejuvenation of Polluted Environment: Microorganisms for Sustainability, vol 25. Singapore: Springer. https://doi.org/10.1007/978-981-15-7447-4_9

Adetunji, C. O., Anani, O. A. and Panpatte, D. (2021). Mechanism of actions involved in sustainable ecorestoration of petroleum hydrocarbons polluted soil by the beneficial microorganism. In: Panpatte, D. G. and Jhala, Y. K. (eds), Microbial Rejuvenation of Polluted Environment: Microorganisms for Sustainability, vol 26. Singapore: Springer. https://doi.org/10.1007/978-981-15-7455-9_8

Alyeksyenko, I., Skyba, E. and Margulov, A. (2019). Biosafety and bioethics are the ground of modern biotechnologies development. *International Journal of Recent Technology and Engineering (IJRTE)* 8(4S): 1–5.

Anani, O. A. and Adetunji, C. O. (2021). Bioremediation of polythene and plastics using beneficial microorganisms. In: Adetunji, C. O., Panpatte, D. G. and Jhala, Y. K. (eds), Microbial Rejuvenation of Polluted Environment: Microorganisms for Sustainability, vol 27. Singapore: Springer. https://doi.org/10.1007/978-981-15-7459-7_13

Anani, O. A., Mishra, R. R., Mishra, P., Enuneku, A. A., Anani, G. A. and Adetunji, C. O. (2020). Effects of toxicant from pesticides on food security: Current developments. In: Mishra, P., Mishra, R. R. and Adetunji, C. O. (eds), Innovations in Food Technology. Singapore: Springer. https://doi.org/10.1007/978-981-15-6121-4_22

Andanda, A. P. (2006). Developing legal regulatory frameworks for modern biotechnology. The possibilities and limits in the case of GMOs. *African Journal of Biotechnology* 5(15): 1360–1369.

Andanda, P. (2009). Status of biotechnology policies in South Africa. *Asian Biotechnology and Development Review* 11(3): 35–47.

Anderson, K. (2010). Globalization's effects on world agricultural trade, 1960–2050. *Philosophical Transactions of the Royal Society of London B, Biological Sciences* 365(1554): 3007–3021. DOI: 10.1098/rstb.2010.0131.

Ayukekbong, J. A., Ntemgwa, M. and Atabe, A. N. (2017). The threat of antimicrobial resistance in developing countries: causes and control strategies. *Antimicrobial Resistance & Infection Control* 6: 47. DOI: 10.1186/s13756-017-0208–x.

Azad, M. S. and Hossan, M. A. (2004). Agricultural biotechnology for green revolution – perceived expectations and potential risks. *Journal of Food Technology* 2(1): 8–17.

Bagheri, A. (2014). Priority setting in Islamic bioethics: top 10 bioethical challenges in Islamic countries. *Asian Bioethics Review* 6(4): 391–401.

Beauchamp, T. and Childress, J. (2013). *Principles of Biomedical Ethics*, New York: Oxford University Press, 7.

Benatar, S. R., Daar, A. S. and Singer, P. A. (2003). Global health ethics: the rationale for mutual caring. *International Affairs* 79: 107–138.

Bhardwaj, M. (1999). *Ethical Issues of Human Genome Project*. Conference on Global Ethos, United Nations University, Tokyo, Japan. Available from: http://vulab.ias.unu.edu/GlobalEthos/papers/minakshi.html.

Blackstone, E. A., Fuhr Jr, J. P. and Steve Pociask, M. A. (2014). The health and economic effects of counterfeit drugs. *American Health Drug Benefits* 7(4): 216–224.

Bobyrov, V. M., Vazhnicha, O. M. and Devyatkina, T. O. (2012). *Basics of Bioethics and Safety*. Nova Knyha.

Bomgartner, M. (2016). The problem with vanilla. *Chemical and Engineering News* 94(36): 38–42. Available from: http://cen. acs.org/articles/94/i36/problem-vanilla.html.

Bracanovic, T. (2012). From integrative bioethics to pseudoscience. *Developing World Bioethics* 12(3): 148–156.

Brooke, A. (2008). *Universal Human Rights in a World of Difference*, 57. Cambridge: Cambridge University Press, p. 388.

Builders, M. I., Builders, P. F. and Iliya, L. (2019). The perception of healthcare providers on the quality of locally made drugs and imported drugs. *Journal of Drugs and Pharmaceutical Science* 3(1): 12–19.

Cascais, A. F. (1997). Bioethics: history, scope, object. *Global Bioethics* 10(1–4): 9–24, DOI: 10.1080/11287462.1997.10800712.

Dwivedi, N., Dwivedi, S. and Adetunji, C.O. (2021) Efficacy of microorganisms in the removal of toxic materials from industrial effluents. In: Adetunji, C.O., Panpatte, D.G. and Jhala, Y.K. (eds), *Microbial Rejuvenation of Polluted Environment. Microorganisms for Sustainability*, vol 27. Singapore: Springer. https://doi.org/10.1007/978-981-15-7459-7_15

Dizon, F., Costa, S., Rock, C., Harris, A., Husk, C. and Mei, J. (2016). Genetically modified (GM) foods and ethical eating. *Journal of Food Science* 81(2): R287–R291. DOI: 10.1111/1750–3841.13191.

Entwistle, V. A., Carter, S. M., Cribb, A. and McCaffery, K. (2010). Supporting patient autonomy: the importance of clinician-patient. *Journal of General Internal Medicine* 25(7): 741–745.

ETC Group (2013). Vanilla and synthetic biology. A case-study. In *ETC Group SynBio Case Studies*. Available from: http://www.ETCgroup.org/files/Vanilla_SynBio_case_study_Oct2013.pdf. Accessed 22 Sept 2017.

FAO (1996). *Rome Declaration on World Food Security*. Available from: www.fao.org.

Fokunang, C. N., Ndikum, V., Tabi, O. Y., Jiofack, R. B., Ngameni, B., Guedje, N. M., Tembe-Fokunang, E. A., Tomkins, P., Barkwan, S., Kechia, F., Asongalem, E., Ngoupayou, J., Torimiro, N. J., Gonsu, K. H., Sielinou, V., Ngadjui, B. T., Angwafor, F 3rd., Nkongmeneck, A., Abena, O. M., Ngogang, J., Asonganyi, T., Colizzi, V., Lohoue, J. and Kamsu-Kom (2011). Traditional medicine: past, present and future research and development prospects and integration in the National Health System of Cameroon. *African Journal of Traditional, Complementary and Alternative Medicines* 8(3): 284–295. DOI: 10.4314/ajtcam.v8i3.65276.

Freemont, P. F. and Kitney, R. I. (2012). *Synthetic Biology*. New Jersey: World Scientific.

Friesen, P., Kearns, L., Redman, B. and Caplan, A. L. (2017). Rethinking the Belmont report? *The American Journal of Bioethics* 17(7): 15–21.

Heemskerk, W., Schallig, H, De Steenhuijsen Piters, B. (2006). *The World of Artemisia in 44 Questions*. The Hague: The Royal Tropical Institute.

Hoberman, J. (2016). Why bioethics has a race problem. *The Hastings Center Report* 46(2): 12–80.

Horne, L. C. (2016). Medical need, equality, and uncertainty. *Bioethics* 30(8): 588–596.

Iaccarino, M. (2001). Science and ethics. *EMBO Reports* 2(9): 747–750. DOI: 10.1093/embo-reports/kve191.

Islam, S., Thangadurai, D., Adetunji, C. O., Nwankwo, W., Kadiri, O., Makinde, S., Michael, O. S., Anani, O. A. and Adetunji, J. B. (2021). Nanomaterials and nanocoatings for alternative antimicrobial therapy. In: Kharissova, O. V., Martínez, L. M. T. and Kharisov, B. I. (eds), *Handbook of Nanomaterials and Nanocomposites for Energy and Environmental Applications*. Cham: Springer. https://doi.org/10.1007/978-3-030-11155-7_3-1

Jefferson, C., Lentzos, F., Marris, C. (2014). Synthetic biology and biosecurity: challenging the "myths". *Frontiers in Public Health* 2: 115. DOI: 10.3389/fpubh.2014.00115.

Karsjens, K. L., Johnson, J. M. (2003). White normativity and subsequent critical race deconstruction of bioethics. *The American Journal of Bioethics* 3(2): 22–30.

Kramer, M. G., Redenbaugh, K. (1994). Commercialization of a tomato with an antisense polygalacturonase gene: the FLAVR SAVR™ tomato story. *Euphytica* 79: 293–297. DOI: 10.1007/BF00022530.

Kuiken, T. (2011). Nanomedicine and ethics: is there anything new or unique?. *Wiley Interdisciplinary Reviews. Nanomedicine and Nanobiotechnology* 3: 111–118.

Mackenzie, D. (2001). Protection racket. *New Scientist*. 21 July 2001.

Maghari, B. M. and Ardekani, A. M. (2011). Genetically modified foods and social concerns. *Avicenna Journal of Medical Biotechnology* 3(3): 109–117.

Marris, C., Jefferson, C. and Lentzos, F. (2014). Negotiating the dynamics of uncomfortable knowledge: the case of dual use and synthetic biology. *Biosocieties* 9(4): 393–420. DOI: 10.1057/biosoc.2014.32.

Martensen, R. (2001). The history of bioethics: an essay review. *Journal of the History of Medicine and Allied Sciences* 56(2): 168–175.

Martin, K., Shilton, K., Smith, J. (2019). Business and the ethical implications of technology: introduction to the symposium. *Journal of Business Ethics* 160: 307–317. DOI: 10.1007/s10551-019-04213-9.

Mautner, M. N. (2009). Life-centred ethics, and the human future in space. *Bioethics* 23(8): 433–440.
Mcguire, A. L., Caulfield, T. and Cho, M. K. (2008). Research ethics and the challenge of whole-genome sequencing. *Nature Reviews Genetics* 9(2): 152–156. DOI: 10.1038/Nrg2302.
Mobasser, S., Jazayeri, M. R., Khazaei, F., Sadeghi, L. (2012). Wheat seed contamination with seed-borne diseases in cold climatic zone of Iran. *International Journal of Plant Production* 6(3): 337–352.
Motoko, A., Kumie, N. and Tetsuya, I. (2014). Bioethics in biotechnology. *Trends in Biotechnology* 32(5): 234–237.
Muldoon, M. and King, N. (1995). Spirituality, healthcare, and bioethics. *Journal of Religion and Health* 34(4): 329–349.
Mishra, P., Mishra, R. R., & Adetunji, C. O. (Eds.). (2020). *Innovations in Food Technology: Current Perspectives and Future Goals*. Springer Nature Singapore Pte Ltd. 2020. Springer, Singapore. Pp. 143-162. 1st Edition. DOI: 10.1007/978-981-15-6121-4
Okan Urker, A., Murat Yildiz, B. and Nesrin Cobanoglu, C. (2012). The role of bioethics on sustainability of environmental education. *Procedia - Social and Behavioral Sciences* 47: 1194–1198.
Peplow, M. (2016) Synthetic biology's first malaria drug meets market resistance. *Nature* 530: 389–390.
Pingali, P. L. (2012). Green revolution: impacts, limits, and the path ahead. *Proceedings of the National Academy of Sciences* 109(31): 12302–12308. DOI: 10.1073/pnas.0912953109.
Polkinghorne, J. C. (2000). Ethical issues in biotechnology. *Trends in Biotechnology* 18: 8–10.
Potter, V. R. (1970). Bioethics: the science of survival. *Perspectives in Biology and Medicine* 14: 127–153.
Potter, V. R. (1971). *Bioethics: A Bridge to the Future*. Englewood Cliffs, NJ: Prentice-Hall, 196p.
Prince Charles HRH (2000). Respect for the earth. *Lecture 6 BBC Reith Lectures 2000*. Available from: http://news.bbc.co.uk/hi/english/static/events/reith_2000/lecture6.stm.
Rastogi, S. and Kaphle, K. (2011). Sustainable traditional medicine: taking the inspirations from ancient veterinary science. *Evidence-Based Complementary and Alternative Medicine* 2011: 151435. DOI: 10.1093/ecam/nen071.
Reich, W. (1995). The word 'bioethics': the struggle over its earliest meanings. *Kennedy Institute of Ethics Journal* 5: 19–34.
Reich, W. T. (1999). The "wider view": André Hellegers's passionate, integrating intellect and the creation of bioethics. *Kennedy Institute of Ethics Journal* 9: 25–51.
Rosie, H. (2004). Bioethics for technology? *Current Opinion in Biotechnology* 15: 250–253.
Rowe, J. D., Amijee, F., Brody, S. D., Wandrey, G. G., Dreyer, C. C. (2012). The globalization of agricultural biotechnology: implications for regulatory compliance, stewardship and stakeholder engagement. In Wozniak, C., McHughen, A. (eds). *Regulation of Agricultural Biotechnology: The United States and Canada*. Dordrecht: Springer. DOI: 10.1007/978-94-007-2156-2_16.
Rui, N. (2014). Medical bioethics. *Encyclopedia of Global Bioethics*: 1–10. DOI: 10.1007/978-3-319–05544-2_44-1.
Sarah, J., Emil, J. and Ramazan, A. (2015). Current state of bioethics relating to biotechnology for engineering education American society for engineering education. ASEE Zone III Conference (Gulf Southwest – Midwest – North Midwest Sections. 1–10.
Sawicka, B., Umachandran BE, K., Skiba, D., Ziarati, P. (2020). Chapter 14 – Plant biotechnology in food security In *Natural Remedies for Pest, Disease and Weed Control*. 163–177. Academic Press, Elsevier. DOI: 10.1016/B978-0-12-819304-4.00014-2.
Saxén, H. (2017). *A Cultural Giant: An Interpretation of Bioethics in Light of its Intellectual and Cultural History*. Tampere: Tampere University Press.

Serageldin, I. (1999). Biotechnology and food security in the 21st century. *Science* 285: 387. DOI: 10.1126/science.285.5426.387.

Shuster, E. (1997). Fifty years later: the significance of the Nuremberg code. *New England Journal of Medicine* 337: 1436–1440. DOI: 10.1056/NEJM199711133372006.

Silverman, E. D. (2004). The 5 most pressing ethical issues in biotech medicine. *Biotechnology Healthcare* 1(6): 41–45, 46.

Subrata, S. (2004). Bioethics and biomaterial research. *Trends in Biomaterials and Artificial Organs* 17(2): 1–3.

Thangadurai, D., Dabire, S. S., Sangeetha, J., Said Al-Tawaha, A. R. M., Adetunji, C. O., Islam, S., Shettar, A. K., David, M., Hospet, R. and Adetunji, J. B. (2021). Greener composites from plant fibers: Preparation, structure, and properties. In: Kharissova, O. V., Martínez, L. M. T. and Kharisov, B. I. (eds), *Handbook of Nanomaterials and Nanocomposites for Energy and Environmental Applications*. Cham: Springer. https://doi.org/10.1007/978-3-030-11155-7_21-1.

Ugbogbo, S. N. and Atu, O. M. (2016). Ethical issues in information technology - a conceptual approach. *Igbinedion University Journal of Accounting* 2: 174.

United Nations (1948). *Universal Declaration of Human Rights*, pp. 1–8. https://www.un.org/en/about-us/universal-declaration-of-human-rights.

Welcome, M. O. (2011). The Nigerian health care system: need for integrating adequate medical intelligence and surveillance systems. *Journal of Pharmacy and Bioallied Sciences* 3(4): 470–478. DOI: 10.4103/0975–7406.90100.

Wight, V., Kaushal, N., Waldfogel, J. and Garfinkel, I. (2014). Understanding the link between poverty and food insecurity among children: does the definition of poverty matter? *Journal of Children and Poverty* 20(1): 1–20. DOI: 10.1080/10796126.2014.891973.

6 Science Communication in Agricultural Biotechnology and Public Engagement with Stakeholders in Nigeria

R.S.M. Gidado, A. Isah, and S.O.O. Iweajunwa
National Biotechnology Development Agency (NABDA)

J.C. Igwe
Kaduna State University

N. Muchiri
African Agricultural Technology Foundation

CONTENTS

6.1 INTRODUCTION

Agricultural Biotechnology is a multidisciplinary approach (Hartley et al., 2016) that involves the application of scientific techniques such as recombinant DNA technology, tissue culture, molecular diagnostics and gene editing technology to modify living organisms including plants, animals and microorganisms with the aim of making or modifying plants genome that will express specific traits such resistance

DOI: 10.1201/9781003179177-6

to diseases, stress tolerance, improved flavour, flower colour and fruit size (Daniel, 2019). Crop biotechnology is one of the many scopes of agricultural biotechnology that has been greatly advanced in recent times. Agriculture generally relies upon science and technology to meet demand in an efficient and sustainable manner.

Though, several renowned scientists have demonstrated that the application of agricultural biotechnology has the potential to transform the livelihoods of the several millions of poor people scattered across the various poverty regions around the globe (Hartley et al., 2016; Navaro, 2011), continuous debate fuelled by scepticism and myths have limited the application of the technology in benefiting lives. Notable socio-economic and environmental benefits can be achieved by several resource-poor farmers in Nigeria through the adoption of crop biotechnology (Edgar et al., 2002).

Modern biotechnology holds a great promise to contribute real socio-economic benefits to the African continent giving her great need for enhanced agricultural productivity. In many African countries including Nigeria, the most populated country in Africa, the problem of food and nutrition insecurity has continued to slow down economic progress. Though the importance of agriculture in driving the economy of Nigeria has been repeatedly underscored by the country's Federal Government, production constraints like soil infertility, insect pest infestations and lack of improved agricultural technologies have continued to dwindle Nigeria's agricultural productivity.

Agricultural Biotechnology has, however, been identified by the Federal Government of Nigeria as a promising technology that has the possibilities of transforming the socio-economic status of Nigeria through its wide range of benefits. The approval of the biosafety guidelines by the Federal Executive Council (FEC) in 2001 with the aim of fast tracking the application of GE crops and the creation of National Biotechnology Development Agency (NABDA) in that same year are clear indices of the government's interest in crop biotechnology. The series of developmental strides undertaken by Nigeria's Federal Government in the area of crop biotechnology might have seemed uncomplicated and hitch free but for the unrelenting anti-Genetically Modified Organisms (GMO) movement spare headed by Greenpeace International and Friends of the Earth. The strong spirited campaigns by these anti-GMO advocacy groups against the adoption of biotechnology application in agriculture through the creation of appropriate policies, proper legal frame work and adequate regulatory environments that will boost the development and commercialization of GMOs in Nigeria have made OFAB-Nigeria to intensify her campaign for the adoption of biotechnological tools in Nigeria's agricultural system. The anti-GMO activists have taken advantage of the poor understanding of agricultural biotechnology among Nigerians and have continued to twist an otherwise scientific and technical issues into religious and ideological debate.

They methodically twist scientific facts by creating alternative stories in the form of myths to distract stakeholders and the general public from seeing any possible hope of the technology having positive impacts on human health, environment and economy of the country (Capalbo and Antle, 2015).

These arguments are often exaggerated by social media, ignorance and pseudoscience with a large fraction of the populace having little or no understanding of the science behind the technology. Therefore, a major challenge for GM crops developing and processing institutions is to determine how to improve the public's perception of the safety and application of GM crops (Solomon et al., 2014).

The need for a more rapid and holistic science communication in Nigeria has never become more necessary than now: with an estimation of 263 million people by 2030 (United Nations, 2015), increasing rate of hidden hunger (Uche and Familusi, 2018), pressure on the government to meet up with the Sustainable Development Goals (SDG) of zero hunger by 2030 (Gil et al., 2019), and the continuous spread of scientific myths (Letrud and Hernes, 2019), science communication that is centred on public enlightenment is now an emergency in Nigeria. Survey conducted by the Open Forum on Agricultural Biotechnology (OFAB) in Africa, Nigeria Chapter, has underscored 'ignorance' as the biggest threat to the advancement of Nigeria using agricultural biotechnology tools. Mutual understanding and dialogue were highlighted by the International Service for the Acquisition of Agri-biotech Applications (2020) as part of the key factors that will drive the understanding of the technology by stakeholders across the world and ensure public acceptance.

The debate on adopting GM crops and animals in Africa requires the application of good communication techniques and appropriate channels that is region targeted and people oriented. The science communicator needs to display high level of confidence and transparency in order to accurately influence and strengthen the opinion and perception of the public regarding the introduction and application of a technological innovation. Strategic communication cannot be an afterthought; it must be given proper investment from the beginning and included in every scientific project budget. A recommendation by the Asia-Pacific Association of Agricultural Research Institutions (APAARI) (2008) reiterated the importance of scientists to be good communicators of their research. By working together in combining stories, evidences, values and effective communication, we can change the public opinion around biotechnology.

Communication means:

1. Information transfer between a sender and a receiver.
2. Social dialogue among and or between a varied audience.
3. Value or believe exchange.

6.2 SCIENCE COMMUNICATION: BUILDING CONSENSUS ON CROP BIOTECHNOLOGY

The communication of science does not only entail the exchange of science products but about making the public to understand and appreciate scientific culture, research results and interpretations that will enable them make informed decisions based on evidence. Achieving trust as a result of mutual understanding that is originated from good dialogue rather than through statements of facts is the goal of science communication (Gregory and Miller, 1998). Hence, communication always requires stakeholders to partake in the communal processes of discussion and decision-making.

Velasco (2005) described the communication of science as creating a common forum where the members of the public, scientists and policy makers meet to dialogue in order to appreciate, accept and understand the new scientific innovations and techniques such as the genetic modification technology. Societal and policy makers engagement during the commercialization stage of emerging technologies has now become a necessity if such scientific innovation must become a success (Einsiedel, 2009).

Jamias (2000) has underscored three outstanding subdomains of science communication to include scientific communication; technical communication that encompasses a semi-popular science communication; and a communal communication of scientific innovations that is public engaging (Jamias, 2000). Different levels of communication spanning from the level of educating to information gathering, discussion and engagement of the audience to partnering have been underscored by the Canada Health (2006) (see Table 6.1).

Science communicators can be scientists, industrialists, academics, media practitioners and farmers and can all be linked by the goal of bridging the gap between science and society. A good science communicator should be able to transform the information within their reach into a form of public knowledge that will result to action. Matsuura (2008) has underscored access to accurate information as a source of empowerment to the public because information is power.

TABLE 6.1
Different Levels and Methods of the Public Involvement Continuum

Level	Type	When Used	Methods
1	Inform or educate	Results should be known by the public from the already-established decision	Fact sheets Press release Mapping of community Social marketing Information kits Campaign on public awareness
2	Information gathering	Shaping of policy decisions ongoing No factual information No opinion information	Stakeholder's meeting Community meetings Focus groups Seminars surveys and public hearings
3	Discussion	Need two-way info exchange Existence of the chance to influence final decision	Bilateral meetings Infotech-based Issue conferences and workshops Technical consultations Constituent assembly Roundtables Citizen's panel
4	Public engagement	Policy directions can be shaped by citizens Citizens discussion on value-laden issues	
5	Partnering	Partnership programme development Citizen's empowerment Development of solutions by citizens	Consensus conference

Source: Canada Health (2006).

6.3 ROLE OF MEDIA

The crucial role played by the media in information dissemination has made it a non-negligible part of science communication. Being the most preferred source of information by members of the public, the media has been identified as an important medium that can help policy makers take a well-informed decision that can create a suitable environment for the application agricultural biotechnology in any society (Scheufele and Tewksbury, 2007). By facilitating dialogue among different sectors of the society, the media enables members of the society to reach a consensus in any scientific matter (Khan, 2006), and media has the capacity to frame the way scientific opinions are portrayed in the public. The concept of framing has been identified by DeVreese (2005) as an act of information re-structuring in such a way that it becomes more meaningful before the public. The type of thematic slant or frame used to write an article such as a new research result can determine the manner in which the public view such a scientific innovation. Brossard et al. (2007) have recommended that biotechnology should be discussed as the perspective of an issue-cycle where attention to the topic is being triggered by varying levels of events.

6.4 THE LAUNCH OF THE OPEN FORUM ON AGRICULTURAL BIOTECHNOLOGY IN AFRICA: AN OVERVIEW

The OFAB was initiated by the African Agricultural Technology Foundation (AATF) under the African Union/New Partnership for Africa's Development (AU/NEPAD) in alliance with congenial organizations in Africa (AATF, 2017). The launch of OFAB as a platform in 2006 was done in order to aid the flow of exchange of biotechnology information among various stakeholders from different background. This became necessary because of the reluctance of most countries in Africa towards the adoption of biotechnology-derived products which have been linked to so many factors, principal among them include unscientific fears and concerns. The confrontation by policy makers in Africa by contrary sources of information have resulted to scientific facts mixed with religious, tribal, political and social considerations. With the confrontation of a rapid population growth accompanied by land fertility that has resulted into low agricultural productivity, policy makers are always enthusiastic to making the right decisions once there is appropriate guidance (Solomon et al., 2014).

For example, following the diverse roles biotechnology could play in the transformation of Africa's economy, the African Union (AU) had to put a high-level modern biotechnology advisory panel that will keep principal officers, most especially, heads of state of different African countries abreast of current trends in biotechnology in order to enable them have a common stand with regard to the acceptance and adoption of various biotechnological innovations. Due to the principal role of AATF in driving the advancement of innovative technologies (one of which is biotechnology) that will boost productivity of plants and animals in Africa, OFAB was established as a platform to enhance the flow of information between the body of scientific and the public stakeholders including the policy makers. The OFAB platform brings various stakeholders of agricultural biotechnology together while facilitating the exchange of information among journalists, lawmakers, scientists, activists and members of various professional bodies in the general public. By serving as a platform by which stakeholders can explore novel ways of integrating biotechnology into the African Agricultural Sector, OFAB encourages research scientists to initiate interactive sessions that is centred on the link between science and socio-economy, health and other human (Solomon et al., 2014). OFAB also manages the superfluous controversy and fears mounted by the anti-GM advocacy groups in the general public by publishing time to time factsheet on the benefits and safety of agricultural biotechnology products.

The media briefing sessions organized by OFAB have become a good source of information gathering for the various stakeholders in the agricultural, social, economic, educational and political sectors. There are Seven Chapters of OFAB in Africa, OFAB-Kenya Chapter domiciled at the International Service for the Acquisition of Agricultural Biotechnology Applications (ISAAA) was the first to be launched in Nairobi, 14 September 2006, followed by OFAB-Uganda in 2007 hosted by Uganda National Council for Science and Technology (UNCST). OFAB-Nigeria and Tanzania Chapters, 2009, came through the partnerships of NABDA and the Commission for Science and Technology (COSTECH), respectively. Other Chapters include: OFAB-Ghana (2011), hosted by the Council for Scientific and Industrial Research-Science and Technology Policy Research Institute (CSIR-STEPRI), OFAB-Burkina Faso, 2012 hosted by the Environmental Institute for Agricultural Research (INERA) and OFAB-Ethiopia, 2014, hosted by the Ethiopian Institute for Agricultural Research (EIAR). With a total of Seven Chapters currently, OFAB Africa is changing the narratives of agricultural biotechnology in Africa and the world at large. The funding of the project is by Bill and Melinda Gates Foundation (BMGF) through AATF. While the Program for Biosafety Systems (PBS); the United States Department of Agriculture (USDA) and the United States Agency for International Development (USAID) also give some funding support (Solomon et al., 2014).

In Nigeria, OFAB is co-hosted by the Agricultural Research Council (ARCN) and managed by a Programming Committee drawn from likeminded institutions such as the National Agricultural Seed Council of Nigeria (NASC); Institute of Agricultural Research (IAR), Zaria; Federal Ministry of Agriculture and Rural Development (FMARD); National Root Crops Research Institute (NRCRI), Umudike; and National Cereals Research Institute, Badeggi.

To efficiently carry out its mandate and achieve her objectives, which include:

i. Establishing and managing various platforms for the easier understanding of biotechnology applications in agricultural productivity;
ii. Informing policy decision-making processes on agricultural biotechnology matters by providing a well robust facts that is backed up with scientific research findings;
iii. Forging strategic alliances that will maximize the utilization of resources by encouraging inter-institutional networking and knowledge sharing in agricultural biotechnology sector;
iv. Aggrandizing a well-targeted capacity building for the improvement of biotech communications across all sectors in Africa.

OFAB-Nigeria has strategically formed a formidable alliance and partnerships with various governmental and non-governmental institutions such as:

1. National Agricultural Seed Council of Nigeria (NASC)
2. Agricultural Research Council of Nigeria (ARCN)
3. Programs for Biosafety System (PBS)
4. National Veterinary Research Institute (NVRI), Vom, Plateau State
5. Federal Ministry of Agriculture and Rural Development (FMARD)
6. States and Local Governments
7. Several professional bodies like the Nigerian Medical Association (NMA)
8. Religious bodies
9. National Root Crops Research Institute (NRCRI), Umudike
10. Institute of Agricultural Research (IAR), Zaria
11. National Cereals Research Institute, Badeggi, Niger State
12. Africa harvest Biotechnology Foundation International (AFBFI)
13. African Seed Trade Association (AFSTA)
14. National Productivity Centre (NPC), Abuja
15. National Institute for Oil Palm Research (NIFOR), Benin, Edo State
16. Cocoa Research Institute (CRIN), Ibadan, Oyo State
17. The Association of General and Private Medical Practitioners (AGPMMP)
18. National Institute of Fresh Water Fisheries (NIFFRR) New Bussa, Niger State
19. The Association of Catholic Medical Practitioners (ACMP)
20. United States Department of Agriculture (USDA)
21. United States' Department of States (USDS)
22. United States Agency for International Development (USAID)
23. Schools (Universities, Secondary) and Professional Bodies
24. Crop Life International
25. Mahyco Agriculture Nigeria Private
26. Bayer Agriculture Nigeria Plc.

Through partnerships and alliances, OFAB-Nigeria Chapter has increased level of awareness and public engagement, bridged the gap between research scientists and various media practitioners; and supportive regulatory environment on agricultural biotechnology. Nigeria provides a strong example of the OFAB model for policy change and policy implementation. For many years, the regulatory environment for genetically modified (GM) technology in Nigeria was partially functional. The country had a Policy on Modern Biotechnology with Biosafety Guidelines and Regulations that only permitted experimentation on GM crops in contained and confined facilities. At that time, the institutional framework for biosafety in Nigeria was a National Biosafety Committee (NBC) whose members worked on ad hoc basis to review applications, while maintaining concurrent full-time research positions in various tertiary institutions. Such regulatory oversight arrangement was ill-suited to oversee environmental release and commercialization of GM crops emerging from the research and development (R&D) pipeline.

Starting in 2009 and launched by the then Hon Ministers of Science and Technology and that of Information and Culture, Dr. Alhassan Zaku and Prof. Dora

FIGURE 6.1 The launch of OFAB-Nigeria on 9 April 2009.

Akunyili (Figure 6.1) all of blessed Memory (Gidado, 2011; Solomon et al., 2014), the OFAB Community in Nigeria raised level of awareness on agricultural biotechnology and how best to manage it for the benefit of Nigeria, with a specific objective of creating champions and supporters in the parliament. It took several years for these efforts to pay off, with the biosafety bill passed by the House of Representatives and Senate for the first time in 2013, but failed to become law because it did not receive the mandatory presidential assent on time and had to be returned to the National Assembly for a fresh round of debate and legislation. In the second round of debate and legislation, the OFAB Community in Nigeria stepped up its efforts through media (television, print and radio) and grassroots campaigns to elevate voices of farmers wanting access to the technology. These fast-track efforts and combined approach contributed to the president's assent in 2015.

Since then, the OFAB Community focused on efforts to ensure effective implementation of the law via the establishment of the National Biosafety Management Agency (NBMA) in 2015 as the National Competent Authority on Biosafety. It also contributed to the development of detailed Biosafety Regulations in 2017. Dr. Rufus Ebegba was appointed the Director General and CEO of NBMA same 2015. Ebegba had previously received capacity strengthening opportunities including benchmarking visits to countries such as Australia, placing him in a position to steer this new agency (see Figure 6.2).

However, these positive developments re-awakened the anti-biotech groups who responded by mounting sustained campaigns against biotech in the country. They called on the government to halt all biotechnology activities and urged the parliament to recall and review the NBMA Act 2015. This made 2016 a difficult and challenging year for OFAB even with the OFAB Community putting up a determined fight. With this, the Chapter mounted a three-pronged campaign strategy covering high-level policy outreach, parliamentary outreach and grassroots community

FIGURE 6.2 Sensitization workshop towards the passage of the biosafety bill.

mobilization to push back the growing anti-biotech onslaught with clearly visible results. The OFAB-Nigeria Chapter went on to record more wins with dismissal, by the Federal High Court in Abuja, of the case filed by Health of Mother Earth Foundation (HOMEF), challenging approval of *Bt.* cotton for commercialization. Their contention was that the decision was made on a Sunday, which is deemed a rest day. The OFAB Community's efforts, with the AATF Legal team and the Cornell Alliance for Science provided training, legal and technical support to the lawyers, resulting in a successful legal strategy that contributed to the 15 August 2018 positive ruling.

Today, the OFAB Community in Nigeria is perhaps the most dynamic OFAB Chapter due to a number of factors foremost being the well-coordinated efforts and involvement of partners in the biotech space such as technology developers, biosafety and regulatory capacity building institutions, and science promoters like the Cornell Alliance for Science that helped cultivate a body of active science advocates through advocacy courses (Solomon et al., 2014).

Other factors included outreach to and on-boarding of high-level influencers such as former Head of State, Chief Olusegun Obasanjo (Figure 6.3) and the then Archbishop of Abuja Archdiocese, His Eminence, Cardinal John Onaiyekan in Nigeria (Figure 6.4); effective utilization of different communication channels including radio and social media, ability to grow the OFAB Community and widen funding sources, e.g., through government agencies such as NABDA and USDA.

However, this progress made by OFAB-Nigeria has not stopped the environmental activists' groups (e.g., the opposition groups to GMOs). They are still active, especially now that *Bt* Cowpea and *Bt* Cotton have entered commercialization phase (Figure 6.5).

Given the recent approvals for cotton and cowpea and the urgency within the opposition to stop this progress, it is therefore apparently critical for the OFAB-Nigeria Chapter to safeguard the gains made so far by ensuring that the enabling environment for the continuous implementation of the biosafety system is kept intact.

Nigeria has progressed its biotech agenda and has shown capability and diligence in handling a variety of GM traits with ease. There has been little high-level interference and the NBMA has been allowed to carry out its mandate effectively. OFAB-Nigeria

FIGURE 6.3 OFAB-Nigeria south-west sensitization workshop with the former President of Nigeria, Chief Olusegun Obasanjo, GCFR, GCON.

FIGURE 6.4 OFAB stakeholders in group photograph with his eminence, John Cardinal Onaiyekan (Middle) after a workshop with the Catholic Priests and Reverend Sisters in Abuja in 2011.

Community has made tremendous progress in just under 10 years, with regulatory approvals of GM crops including a GM food, the pod-borer-resistant (PBR) Cowpea. The OFAB Community is, however, weary of any possible interference especially

FIGURE 6.5 World press conference for the commercialization of the *Bt.* Cotton.

with change of government and/or office holders and will therefore work towards consolidating the gains made, safeguarding the system and ensuring successful commercialization and use of the crops to build confidence in the technology. The Community will also support realization of an efficient and quality delivery system that will build sustainable the uptake and utilization of the products released to date – *Bt* cotton and PBR Cowpea. This will entail working with value chain actors such as seed producers/marketers and farmers.

The OFAB-Nigeria Team will therefore work towards widening community support specifically among professional bodies including the business community, faith-based organizations, medical practitioners, nutrition society and legal fraternity that will help build local ownership and sustain the market. It will build a strong information and education network that will on-board the government extension services that has a structured network of reaching farmers; medical and nutrition practitioners as expert advisors on the technology; and include more grassroots communities especially women, youth and farmer influencer leaderships in promoting and defending the technology and products. The chapter will continue with proactive management of the opposition to reduce any negative impacts on progress being made (Solomon et al., 2014).

6.5 HIGHLIGHTS OF OFAB-NIGERIA ACTIVITIES

Laudable advancements have been achieved by OFAB-Nigeria in the establishment and management of wide range of platforms for the enhancement knowledge sharing, awareness creation, improved understanding and appreciation of modern biotechnology application in agriculture (Figure 6.6).

FIGURE 6.6 Highlights of key activities of OFAB-Nigeria in 2018 that contributed to her progress.

6.6 PUBLIC ENGAGEMENTS: INCREASING UNDERSTANDING OF MODERN BIOTECHNOLOGY

A three-pronged approach was developed by OFAB-Nigeria that played significant roles to the advancement of awareness creation, knowledge sharing and exchange of information and experiences (Figure 6.7).

With the methods of using country side sessions, the platform has held meetings targeting farmers, researchers and civil society groups and between 2012 and 2020, OFAB conducted workshops (Figure 6.3) in partnership with States and other likeminded institutions (public/private) and professional bodies covering 19 out of 36 States of the Federation, Participants (ranging from 100 to 500 at each session) included State leaders, farmer groups, the media, civil society organizations and consumer groups (Figure 6.8).

These sessions covered many states in all the six geopolitical zones including: North-East, North-West, North-Central, South-East, South-West and South-South (Figure 6.9). The third strategy encompasses special sessions for discussions between media practitioners, policy makers, executive members of various civil society groups and other stakeholders.

In addition to the workshops, OFAB-Nigeria Chapter effectively organizes field tours for highly influential journalists, leaders of the various farmers association in Nigeria, policy makers, regulators and other stakeholders (Figures 6.10 and 6.11) with the aim of making them have a first-hand experience of GM crops, understand the science of genetic modification while appreciating the magnitude of efforts that is required for the development GM crops.

The development of handy information materials in the form of policy briefs, fliers, newsletter, etc. has also facilitated the awareness and understanding of agricultural biotechnology in the country (Figures 6.12 and 6.13). Most of these information materials created by OFAB-Nigeria now serve as a resource material for

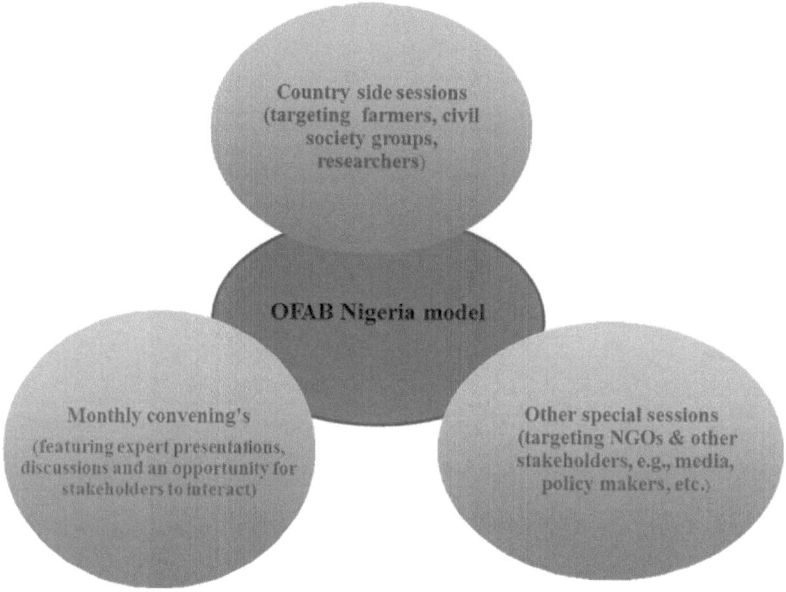

FIGURE 6.7 OFAB-Nigeria three-pronged approach.

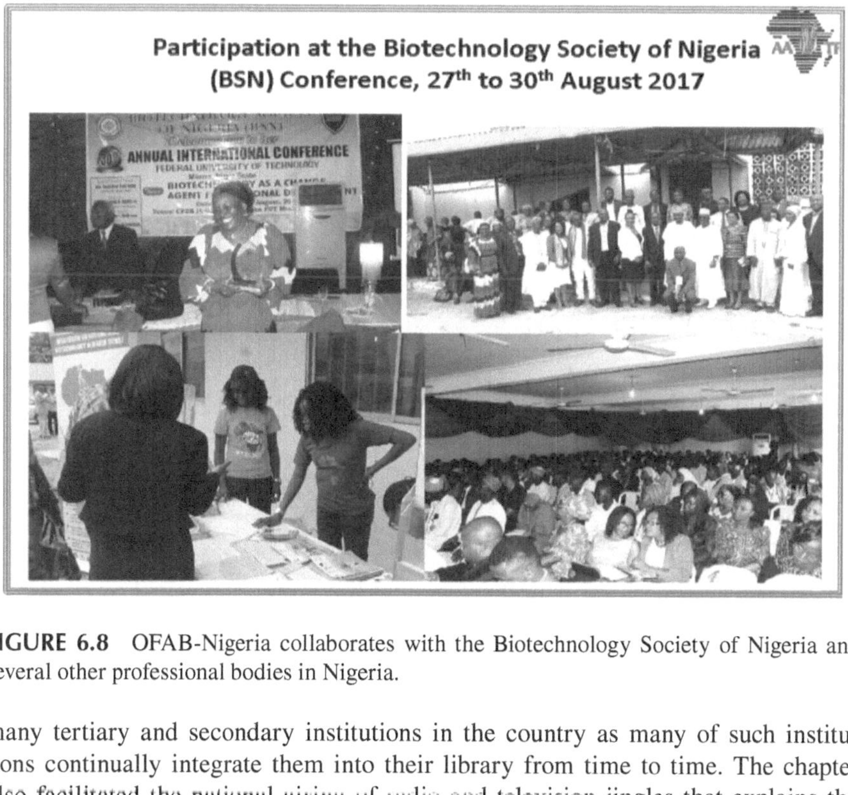

FIGURE 6.8 OFAB-Nigeria collaborates with the Biotechnology Society of Nigeria and several other professional bodies in Nigeria.

many tertiary and secondary institutions in the country as many of such institutions continually integrate them into their library from time to time. The chapter also facilitated the national airing of radio and television jingles that explains the

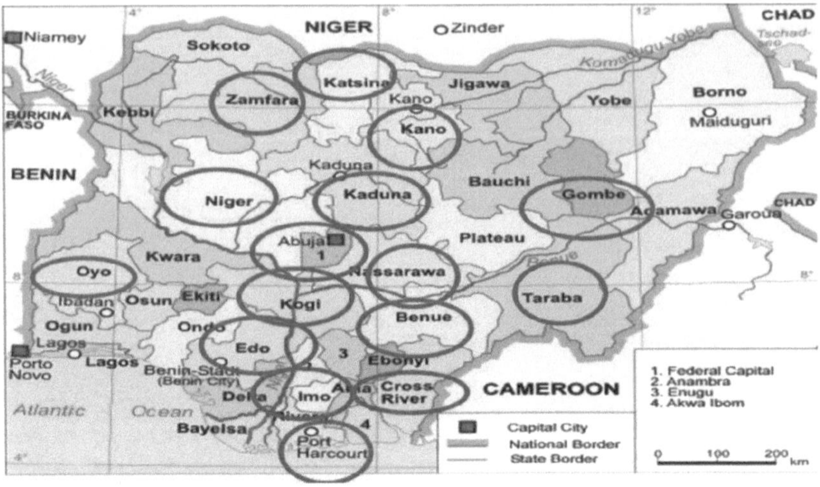

FIGURE 6.9 Target regions of OFAB-Nigeria (Solomon et al., 2014).

FIGURE 6.10 One of the 'Seeing is Believing' tour for law makers to the United States of America in 2018 organized by OFAB-Nigeria.

economic importance and safety of GM crops. Various documentaries that explicitly explain the science of genetic modification of the pod borer resistant GM cowpea; Nitrogen use Efficient, Water use Efficient and Salt Tolerant (NEWEST) Rice and Africa Biofortified Sorghum (ABS) were also created by the OFAB-Nigeria Chapter and are played in major television and radio stations across the country.

The awareness and education campaigns also targeted schools (Figure 6.14).

The inauguration of the Biotechnology and Biosafety Club was done based on the fact that the earlier the children are exposed to agricultural biotechnology, the more their interest can be secured in the technology, and since they are the leaders of tomorrow, they will become pioneers of agricultural biotechnology in the future. Exercise books, biros, t-shirts, almanac and handkerchiefs carrying simplified messages on agricultural biotechnology were also distributed among these students to facilitate easy uptake (Figure 6.15).

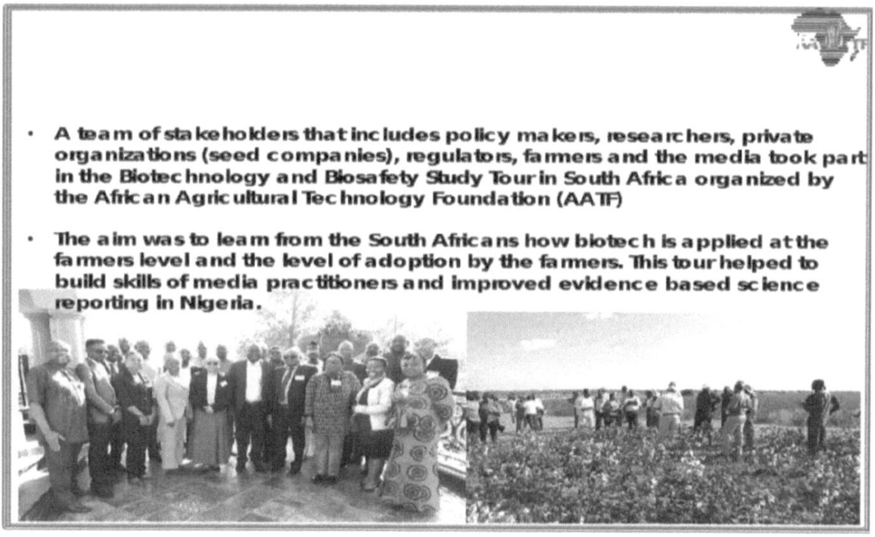

FIGURE 6.11 Biotechnology and biosafety tour to South Africa by policy makers, researchers, seed companies, regulators, farmers and the media.

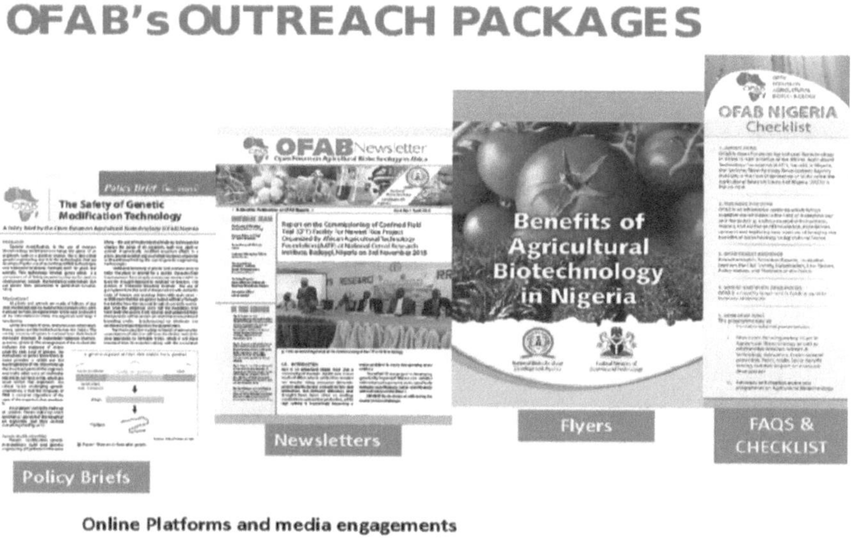

FIGURE 6.12 Some of the many outreach materials by OFAB-Nigeria.

6.7 EFFECTIVE MEDIA ENGAGEMENT

One of the strengths of OFAB-Nigeria is its ability to effectively engage the media to change the reporting on agricultural biotechnology from biased, inaccurate and sensational stories to a more evidence-based and factual stories. This led to an unprecedented change in the narrative of agricultural biotechnology reporting in national newspapers, television and radio stations.

FIGURE 6.13 Some of the outreaches of OFAB-Nigeria to primary and secondary schools.

FIGURE 6.14 OFAB-Nigeria inaugurates biotechnology and biosafety club in primary and secondary schools.

OAFB Nigeria has consistently conducted annual media trainings, workshops, in-house discussions and Gala night for the rewarding of excellence on journalism with the goal of increasing the scope of knowledge and understanding of media practitioners on the diverse application of modern biotechnology in agriculture (Figure 6.11). In addition, the OFAB-Nigeria Chapter has been conducting media tours to the various Confined Field Trials (CFTs) of GE crops (Figure 6.11) which all formed part of

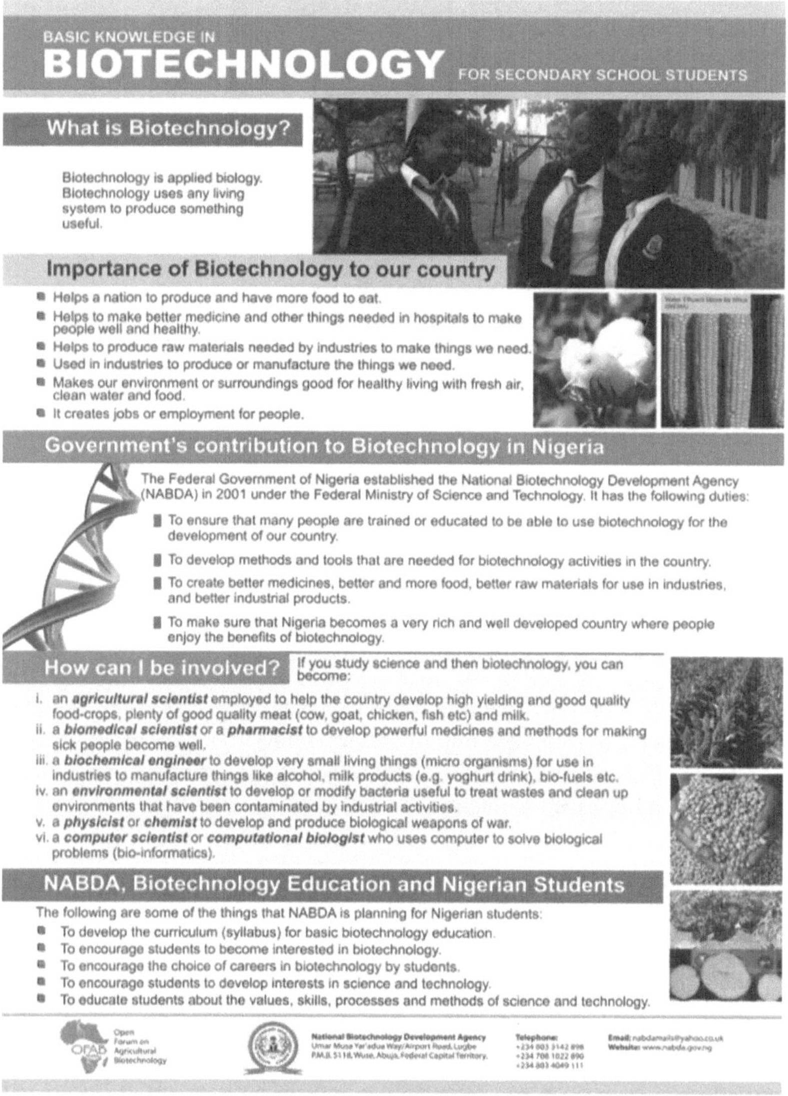

FIGURE 6.15 Biotechnology almanac for high school students.

the bedrock for a fostered understanding and easy communication of agricultural biotechnology application by the media. The quarterly media tracking, monitoring and analysis conducted by OFAB-Nigeria Chapter to assess the performance of the various media houses towards the advancement of agricultural biotechnology application serves as a guiding tool for planning and re-strategizing (Figure 6.16).

The various media trainings conducted by OFAB-Nigeria has enabled journalists to know the difference between the pseudoscience-driven activities of the anti-GMO and the scientific facts and findings obtained as a result of thorough research by scientists. Where in doubt, journalists now take advantage of the OFAB platform to

FIGURE 6.16 OFAB-Nigeria reward for excellence in science reporting.

gain open access to credible scientists to verify some facts that seems not clear to them as many scientists who were initially reserved towards the sharing the benefits of their research experience and outcome with journalists are now open to inquiries from journalists at any given moment.

6.8 INFLUENCING POLICY AND REGULATORY ENVIRONMENTS

At the time when the launching of OFAB-Nigeria took place in 2009, there were little or no possibility that the uncompleted biosafety bill will successfully be passed into law after undergoing several antagonism and delays for seven year since conception. However, the strategic partnership formed by OFAB-Nigeria with likeminded institutions like NABDA, ARCN, Biosafety Desk Office of the Federal Ministry of Environment, the African Biosafety Network of Expertise (ABNE), PBS, Africa Harvest Biotechnology Foundation International and many other research institutes in Nigeria, enabled the chapter to accelerate the passage of the biosafety bill.

The unstinting and assiduous efforts of OFAB-Nigeria and her partners eventually became rewarding on 18th April 2015 when the then sitting President, Goodluck Ebele Jonathan and the 7th National Assembly signed and passed the bill into law (Gidado, 2012).

Annual conference meetings, workshop and symposia conducted by the various professional bodies including the Biotechnology Society of Nigeria (BSN); Nigerian Medical Association (NMA); Nigerian Bar Association (NBA); National Council on Science and Technology (NCST); the Catholic Secretariat of Nigeria engaging Catholic Priests, Rev. Sisters, Medical Doctors; the Association of Catholic Medical practitioners of Nigeria; Genetic Society of Nigeria (GSN); Nigerian Seed Companies

FIGURE 6.17 Special engagements.

were keyed into by OFAB-Nigeria, using that as a platform to educate her members and executive on the science, safety and benefits of genetic modification and products which resulted into improve public understanding and acceptance of GM crops. The strategic and timely participation of OFAB-Nigeria in one of the meetings of the National Agricultural Council resulted to the integration of agricultural biotechnology and biosafety into the National Agricultural Policy Plan (2016–2020) of Nigeria.

In all of these strategies by the chapter, the youth were not left out as OFAB-Nigeria co-organized football competition for student of Nasarawa State University Keffi with the aim of getting their attention for the awareness on GM crops application, safety and benefits to the Nigerian economy via employment creation. Interactions on the benefits, safety and application of the genetic modification technology and products (GM crops) were made during the break after the halftime of the football match. OFAB-Nigeria crested vest with short agricultural biotechnology campaign message and OFAB web address were presented to the winning team (see Figure 6.17).

In addition to some of the routines of OFAB-Nigeria in the form of advocacy visits to media houses (Radio, Television, Newspapers and magazines), one-on-one engagement with policy makers, farmers, media, civil societies, etc., the use of social media platforms (Facebook, Twitter, LinkedIn, Instagram, YouTube and Tumblr) has also played significant roles in OFAB Giant Advocacy strides. The Chapter has continued to gain massive inflow of followers due to her active postings on a daily basis. It has also used this medium to keep members of the general public abreast with latest information and activities regarding agricultural biotechnology research all over the world, including strong statements from Nigerian policy makers (Figure 6.18). Livestreaming of most OFAB activities are also done to enable the public follow-up while posting their comments and questions for instant response.

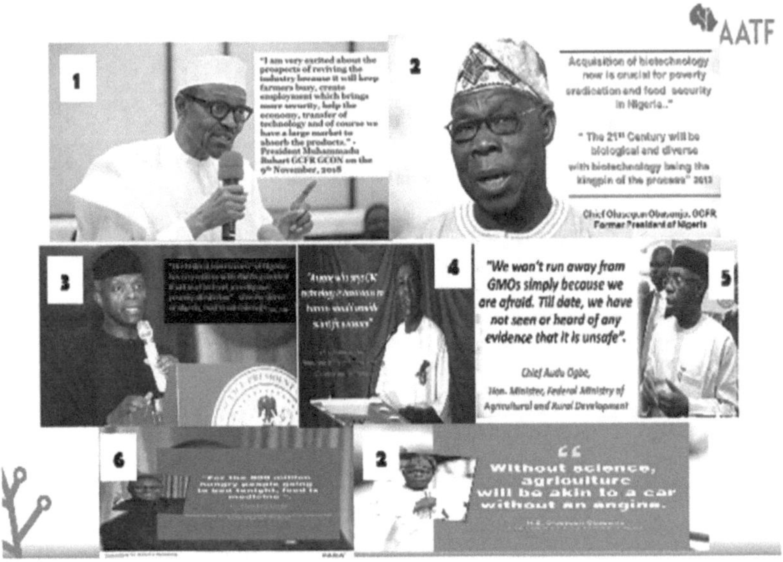

FIGURE 6.18 Quotes by Nigerian Policy Makers including: 1. the current Presidents, His Excellency, General Muhammadu Buhari; 2. the former President of Nigeria, Chief Olusegun Obasanjo; 3. the current Vice President, Prof. Yemi Osinbajo (SAN); 4. Current Minister of Science and Technology, Hon. Ogbonnaya Onu; 5. Former Minister of Agriculture and Rural Development, Chief Audu Ogbe; 6. Executive Director, Dr Yemi Akinbamijo.

You can follow OFAB-Nigeria via:

- **Facebook:** Open Forum on Agricultural Biotechnology in Africa, Nigeria Chapter
- **Twitter:** @OFABnigeria
- **Linkedin:** www.linkedin.com/in/ofabnigeria
- **Instagram:** https://instagram.com/ofab_nigeria
- **YouTube:** OFAB-Nigeria
- **Tumbr:** ofabnigeria.tumblr.com

6.9 CONCLUSION

Attention and priority misplacement on other variables (such as fear, myths and sentiments) other than science has been the sole reasons for the continuous fuelled debates surrounding safety and benefits of agricultural biotechnology in our contemporary world. The polarization of stakeholders due to conflicting and pseudoscience messages has created the presence of divergent information sources, among others, slowing the advancement of the biotechnology in Africa's agricultural system. The need for science communication for the creation of an enabling environment where the practice and application of crop biotechnology is encouraged is now an emergency. Achieving a great success in the warding-off of myths and gimmicks by the anti-GMO advocacy group requires a concerted effort of stakeholders of diverse origins. In Nigeria already, policy makers are beginning to become champions and advocates of crop biotechnology. High top ranked government officials such as the Minister of Science and Technology, Dr Ogbonnaya Onu has come out publicly to declare his stance for biotechnology, describing the biotechnology as the hope of the future of agriculture in Nigeria. It is therefore evident that the more advocacy effort is intensified in Africa, and the more agricultural biotechnology application will be appreciated and adopted. Therefore, more support should be given to science communication in Africa and Nigeria in particular.

REFERENCES

African Agricultural Technology Foundation (2017). *The Open Forum on Agricultural Biotechnology (OFAB): A Decade of Success, 2006–2016*. Nairobi, Kenya: African Agricultural Technology Foundation.

Agricultural Biotechnology: What Are the Issues? The Land Grant University Brochure, Cornell University, College of Agriculture and Life Sciences. Adapted from: http://www.nysaes.cornell.edu/comm/gmo/PDF/GMOLandGrant.pdf.

Asia-Pacific Association of Agricultural Research Institutions (APAARI) (2008). Asia-Pacific Consortium on Agricultural Biotechnology (APCoAB), and Malaysian Agricultural Research and Development Institute (MARDI). Proceedings: Expert Consultation on Agricultural Biotechnology for Promoting Food Security in Developing Countries. Kuala Lumpur, Malaysia. 22 pp.

Brossard, D., James, S. and Clint, N. (Eds.). (2007). *The Public, the Media and Agricultural Biotechnology*. Wallingford, Oxon, United Kingdom: CAB International. 405 pp.

Canada Health (2006). *The Health Canada Policy Toolkit for Public Involvement in Decision Making*. Cat. No.: H39–549/2000E. ISBN: 0–662–29243-X. Accessed on 27th February, 2021. Adapted from: https://www.canada.ca/en/health-canada/corporate/about-health-canada/reports-publications/health-canada-policy-toolkit-public-involvement-decision-making.html.

Capalbo, S.M. and Antle, J.M. (2015). *Agricultural Productivity Measurement and Explanation*. Abingdon, Oxon and New York, NY: Routledge. ISBN 13: 91–31578-1–138–94114-4 (hbk); ISBN 13: 978-1-315–67331-8 (cbk).

Daniel, L. (2019). *What Is Agricultural Biotechnology? Mixer Direct*. Adapted from: https://www.mixerdirect.com/blogs/mixer-direct-blog/what-is-agricultural-biotechnology.

DeVreese, C. (2005). News framing: theory and typology. *Information Design Journal + Document Design*. Vol. 23, No. 1, pp. 51–62. John Benjamins Publishing Co.

Edgar, J., Elias, B. and Adnan, B. (2002). Biotechnology and the developing world. *Electronic Journal of Biotechnology*. Vol. 5, No. 1, Issue of April 15, 2002.

Einsiedel, E. (2009). Introduction: making sense of emerging technologies. In Edna F. Einsiedel (ed.), *First Impressions: Understanding Public Views on Emerging Technologies*. Genome Prairie GE LS Team. University of Calgary, Canada. pp. 2–5.

Gidado, R.S.M. (Ed.). (2011). *Open Forum on Agricultural Biotechnology (OFAB) 2011 Report*. National Biotechnology Development Agency (Federal Ministry of Science and Technology). ISBN: 978-978-923-467-7.

Gidado, R.S.M. (Ed.). (2012). *Open Forum on Agricultural Biotechnology (OFAB) 2011 Report*. National Biotechnology Development Agency (Federal Ministry of Science and Technology). ISBN: 978-978-931-851-3.

Gil, J.D.B., Reidsma, P. and Giller, K. (2019). Sustainable development goal 2: Improved targets and indicators for agriculture and food security. *Ambio Journal*, Vol. 48, 685–698. DOI: 10.1007/s13280-018-1101-4.

Gregory, J. and Miller, S. (1998). *Science in Public: Communication, Culture and Credibility*. Basic Books, 2000. ISBN: 0465024505, 9780465024506, 304 Pages.

Hartley, S., Gillund, F., van Hove, L. and Wickson, F. (2016) Essential features of responsible governance of agricultural biotechnology. *PLoS Biology*, Vol. 14, No. 5, p. e1002453. DOI: 10.1371/journal.pbio.1002453.

International Service for the Acquisition of Agri-Biotech Applications (2020). *Communicating Crop Biotechnology*. Pocket K No. 33. Ithaca, NY: ISAAA. Adapted from: https://www.isaaa.org/resources/publications/pocketk/document/Doc-Pocket%20K33.pdf

Jamias, J. (2000). Towards a definition of science communication. In Proceedings on Discoursing Science Communication: First Roundtable Discussion on Science Communication. Los Baños, Laguna, Philippines: College of Development Communication, University of the Philippines. pp. 27–32.

Khan, M.H. (2006). Governance and development. Paper presented at the 'Workshop on Governance and Development' Organized by the World Bank and DFID in Dhaka, 11–12 November, 2006.

Letrud, K. and Hernes, S. (2019) Affirmative citation bias in scientific myth debunking: a three-in-one case study. *PLoS One*, Vol. 14, No. 9, p. e0222213. DOI: 10.1371/journal.pone.0222213.

Matsuura, M. (2008). Performance Impact of Task-to-Task Communication Protocol in Network-on-Chip: Proceedings of the Fifth International Conference on Information Technology: New Generations. DOI: /10.1109/ITNG.2008.109.

Navaro, M.J. (2011). *Communication Challenges and Convergence in Crop Biotechnology*. Ithaca, New York: International Service for the Acquisition of Agri-biotech Applications (ISAAA) and Philippines: SEAMEO Southeast Asian and Regional Center for Graduate Study and Research in Agriculture (SEARCA). Chapter 2, pp. 26–42.

Scheufele, D.A., and Tewksbury, D. (2007). Framing, agenda setting, and priming: the evolution of three media effects models. *Journal of Communication*, Vol. 57, No. 1, pp. 9–20. DOI: 10.1111/j.0021-9916.2007.00326.x.

Solomon B.O., Ebegba, R., and Gidado, R.S.M. (2014). Influencing politicians and policy makers for a viable biotechnology sector: a case study of the Nigerian biosafety bill drafting, and passage process at the parliament. In F. Wambugu and D. Kamanga (eds.), Biotechnology in Africa, Science Policy Reports 7, Chapter 16. DOI: 10.1007/978-3-319-04001-1_16.

Uche, C. and Familusi, L.C. (2018). Topical review of hidden hunger and poverty profile in Nigeria: case study of Kwara state (Province) and suggested agricultural and economic solutions. *Global Journal of Management and Business*, Vol. 18, No. 4, Version 1.0.

United Nations (2015). *Population 2030: Demographic challenges and opportunities for sustainable development planning (ST/ESA/SER.A/389)*. Adapted from: https://www.un.org/en/development/desa/population/publications/pdf/trends/Population2030.pdf.

Velasco, M.T. (2005). Science communication in the context of development communication: a historical perspective. In *Communication of Scientific and Technical Information*. M.T. Velasco, J. Sison, A. Moran, and A. Flor (Eds.). Los Baños, Laguna, Philippines: University of the Philippines Open University. pp. 269–286.

7 Biotechnology Development in Nigeria

Meeting Critical Developmental Needs

A. Akpa and N. C. Ezebuiro
National Biotechnology Development Agency (NABDA)

Benjamin Ewa Ubi
Ebonyi State University
Tottori University

Christie Onyia
Godfrey Okoye University

Abdulrazak Ibrahim
Ahmadu Bello University

CONTENTS

DOI: 10.1201/9781003179177-7

7.1 INTRODUCTION

Biotechnology is a broad area of biology, involving the use of living systems and organisms to develop or make products and the product of its application depends on the regulatory environment, the needs to be met, and expertise of the handler (Stewart, 2016). Biotechniques such as tissue culture, genetic engineering, molecular breeding, proteomics, genomics, metabolomics, genome editing, and biocatalysis are responsible for the new wave of products and services in the health, agricultural, environmental, and industrial sectors. It suffices to say that the 21st-century global economic winners will be countries with very strong biotechnological capabilities. This underscores the significance of a national framework for coordinating biotechnology research and development, adaptation, and deployment to achieve national strategic targets. This strategic framework must integrate systems thinking, in which the development of each technology meets a corresponding articulated demand, galvanizing all relevant stakeholders in a value chain approach. In recognition of the importance of biotechnology to national development, the Federal Executive Council (FEC) of Nigeria on 23 April 2001 approved the National Biotechnology Policy, which led to the establishment of the National Biotechnology Development Agency (NABDA) in November 2001. Established as a Federal Government Agency under the aegis of the Federal Ministry of Science and Technology, NABDA is responsible for implementing the biotechnology development policy, aimed at promoting, coordinating, and setting agenda for research, development, and adaptation of biotechnology in Nigeria.

7.1.1 Overall Mandate of the National Biotechnology Development Agency

NABDA has the mandate of promoting, coordinating, adapting, and deploying cutting-edge biotechnology research and development, processes, and products for the socioeconomic well-being of the nation. The overarching mission of the Agency is implemented through specific mandates that directly meet national aspirations in the different sectors of the economy and these include inter alia (according to the National Biotechnology Policy, 2001):

i. Carry out well-focused research and development in biotechnology in priority areas of food and agriculture, health, industry, environment, and other strategic sectors for national development and draw up programs and policies for biotechnology utilization, research, and deployment in Nigeria;
ii. Facilitate holistic capacity development in all aspects of biotechnology required for the implementation of the national biotechnology program;
iii. Encourage private sector participation in the biotechnology industry and provide a blueprint for such involvement;
iv. Develop resource mobilization strategy and ensure sustainable funding mechanism for biotechnology activities by engaging national and international donor agencies;
v. Establish and supervise relevant research and development centers for purpose of executing the national biotechnology programs;
vi. Create public awareness and participation in biotechnology development activities through strong advocacy programs, seminars, conferences, and workshop;
vii. Collaborate on biotechnology with international research centers, nongovernmental organizations, industries, other national and international agencies and institutions;
viii. Stimulate biotechnology entrepreneurship schemes to effect rapid commercialization of biotechnology research and products;
ix. Collaborate with and provide support to universities and other academic institutions in the country for research and development projects relevant to the national biotechnology program;
x. Promote sustainable utilization of biological resources and develop novel products to improve the production of our indigenous plants and animals;
xi. Coordinate all biotechnology activities and ensure easy access to stakeholders in matters relating to bioresources and biotechnology;
xii. Serve as national bioinformatics and biotechnology data centers;
xiii. Be the national focal point for bioethics to which Nigeria subscribes; and
xiv. Undertake such other activities as are necessary or expedient for the carrying out of the functions of the Agency and promotion of biotechnology in Nigeria.

7.1.2 Organizational Structure of the National Biotechnology Development Agency

Biotechnology administration in Nigeria is hinged on the administrative structure of NABDA in close working relationship with the National Biosafety Management Agency (NBMA) for regulation. There are several departments and agencies of government that have mandates in different aspects of biotechnology, predominantly in the health and agricultural sectors. Examples include the several national agricultural research systems (NARs) such as the Institute for Agricultural Research (IAR), Ahmadu Bello University, Zaria; the National Root Crops Research Institute (NRCRI), Umudike, the National Cereal Crops Research Institute, Baddeggi; the Cocoa Research Institutes of Nigeria, Ibadan; the Lake Chad Research Institute of Nigeria; the Institute of Agricultural Research and Training, Obafemi Awolowo University, Moor Plantation, Ibadan; the National Institute for Horticultural Research and Training, Ibadan; the Forestry Research Institute of Nigeria, Jericho, Ibadan; the Rubber Research Institute of Nigeria, Benin City; the National Institute for Oil Palm Research; the National Centre for Genetic Resources and Biotechnology, Moor Plantation, Ibadan; Federal Institute of Industrial Research, Oshodi, Lagos; Nigerian Institute of Medical Research, Lagos, etc.

Regarding biotechnology development and deployment, the activities of NABDA are hinged on a governance and operational structure that has visibility across the country. With over 30 Bioresources Development Centres (BIODECs), six Centres of Excellence in the premier universities (University of Nigeria, Nsukka; University of Ibadan; Ahmadu Bello University; University of Maiduguri; University of Port-Harcourt; University of Jos) and an operational headquarters in Abuja, NABDA has an elaborate presence across the country. This enables rapid deployment of biotechnologies for national development across health, agriculture, industry, and environment. Figure 7.1 provides an overview of the structure of biotechnology governance in Nigeria under NABDA.

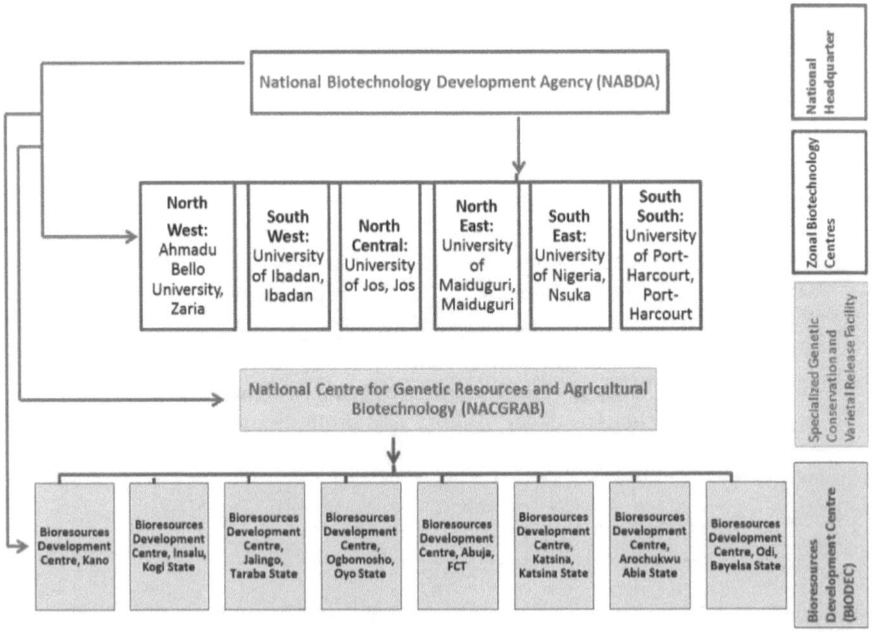

FIGURE 7.1 Overview of the structure of biotechnology governance in Nigeria under NABDA (there are over 30 BIODECs in Nigeria).

7.1.3 Critical Developmental Needs in Nigeria

Nigeria faces critical challenges in the sectors of health, agriculture, industry, and environment. Addressing these challenges especially in the context of food and nutrition security as well as availability of agro-raw material requires concerted efforts predicated on science-policy nexus, to generate the relevant innovations and technologies. In the industrial sector, this need is best underscored by the now comatose textile mills in Nigeria that depended on cotton for the production of textile. The inability of the Nigerian farmers to meet the industrial cotton lint demand resulted in the closure of Nigeria's textile mills with its attendant job losses and revenue to the country.

In the health sector, most of the essential drugs, vaccines, rapid diagnostic kits, and other needed biologicals are derived from biotechnological techniques. However, such techniques are still at their infancy in the country; strengthening Nigeria's capacity to adapt the techniques to meet national demand is a function of the availability of critical mass of human, capital and material resources, policy framework, and institutional arrangements to drive the governance and implementation of the policies. In this respect, NABDA prioritizes four developmental areas: industry, agriculture, health, and environment.

7.2 ADAPTATION AND DEPLOYMENT OF BIOTECHNOLOGY IN NIGERIA

One of the cardinal roles of NABDA is setting agenda for the deployment of biotechnology in Nigeria across all sectors. Although previous emphases focused on developing the critical human and material resources in these sectors, recent efforts have been directed toward commercialization of developed technologies. With some level of competency attained and owing to the increased advocacy and agenda setting activities of NABDA and the biotechnology Society in the Country (the Biotechnology Society of Nigeria (BSN)), there has been relatively increased investment in agricultural, medical, environmental, and industrial biotechnology over the last decade. Presently, the focus of biotechnology development in Nigeria has advanced to the level of technology deployment at commercial scale to meet national needs. Some of the projects and programs are discussed below.

7.2.1 Commercial Release of *Bt.* Crops

Two genetically engineered (GE) crops, *Bt.* cotton and *Bt.* cowpea, with enhanced productivity and insect pest resistance have been released for commercial production in Nigeria.

7.2.1.1 *Bt.* Cotton

Cotton is an important crop due to the reliance of the textile industry on cotton lint, and ginneries on its seed. For decades, cotton production (lint and cotton seed) has been a driving force for economic development in Africa, especially Nigeria (Moseley and Gray, 2008). In Nigeria, the neglect of the agricultural sector during the oil boom years (1970–1980s) negatively impacted on the cotton sector. With surplus revenue from oil sales, the textile industry took the backstage, compromising the entire cotton sector. In 1954, cotton production in Nigeria was 30,000 metric ton (MT) which increased to 60,000 MT in 1979 (http://www.cottontapafrica.org/nigeria.html). Production decline started in the early 1980s and reached 30,000 MT in 1988. In 2018, national production of cotton stood at 94,000 MT (www.fao.org/faostat/en/#data/QC).

The main feature of Nigerian cotton cultivation is that 80% of total production is by small-holder farmers, under rain-fed conditions and with simple tools. Currently, the yield in Nigeria is around 1 ton of seed cotton/ha. Saw ginning is prevalent and the most common Nigerian variety produces fiber of 28–30 mm in length. Cotton in Nigeria has strong linkages with the domestic industries which consume up to 50% of the total production and the rest is exported to EU, China, South Korea, and Taiwan. Nigeria had well-developed textile industries till the 1980s and these were some of the finest and most vibrant industries in the country. The industry started to decline after 2,000 followed by closure of the major factories. To compound the problem of dwindling potentials of what remained of Nigeria's cotton industry, small-holder farmers, whose livelihoods depended on its cultivation, regularly experienced losses due to pests and diseases.

Notable pests of cotton in Nigeria are the lepidopteran insect pest such as the cotton bollworm (*Helicoverpa armigera* Hubn), spiny bollworms (*Earias insulana* Boisd and *E. biplaga* Wlk.), and leafroller (*Sylepta derogata* F.) (Fadare and Amusa, 2003). Of these, the cotton bollworm poses the most significant threat to cotton industry in Nigeria. The insect attacks cotton blooms causing their premature opening without fruits. Most damaged bolls abort while others fail to produce lint or when they do, are of inferior quality. Cotton bollworm also leads to secondary infections by fungi and bacteria and eventually to rotting of fruits. Unfortunately, the control measures, which include the use of insecticides, are not always environmentally friendly and sustainable. These factors have led to the near-comatose situation of the cotton industry in Nigeria.

The Government of Nigeria has given priority to reviving, the once flourishing cotton textile industry. In an effort to revive the entire value chain, with focus on value addition, the Nigerian government has mandated some relevant institutions, including the Manufacturer's Association of Nigeria (MAN), Cotton Development Committee (CDC), and Nigerian Textile Manufacturers Association (NTMA) to implement programs geared toward addressing the challenges facing the textile and garment industries. Given the need to sustain cotton production to meet the challenges of reviving textile and garment industry in Nigeria, NABDA facilitated the deployment of *Bt.* Cotton to Nigeria in 2016 by facilitating the entry of Bayer Agriculture Nigeria Ltd. Commencing work in Nigeria, Bayer Agriculture Nigeria Ltd, IAR, Ahmadu Bello University, Zaria, and NABDA filed an application for the environmental release of *Bt.* Cotton to the NBMA. This was approved. Consequently, the *Bt.* Cotton seeds were deployed into the country and performance testing were carried out for 2 years under the supervision of the National Agricultural Seed Council (NASC) of Nigeria and the National Varietal Release Committee (NVRC). NVRC is a multidisciplinary and multi-stakeholders Committee with the mandate to release improved seeds into the Nigerian seed system.

FIGURE 7.2 (a) Cotton farm in NABDA for the validation of transformation claims of the MRC 7377 BG 11 and MRC 7361 BG 11 varieties commercially released in Nigeria (Mahyco Bollgard II Hybrid Cotton Farm at NABDA); (b) Alhaji Salmanu Abdullahi, Chairman, Cotton Ginners Association sharing his experience on *Bt.* Cotton Farming at a farmer's field day held in NABDA.

The success of the two-year field performance of the *Bt.* Cotton with potential yield of 4.1–4.4 t/ha led to the approval for commercialization of two new varieties of cotton, namely MRC 7377 BG 11 and MRC 7361 BG 11 by NVRC in July 2018. The hybrid cotton varieties were genetically modified to include a gene from *Bacillus thuringiensis* (*Bt.*) which confers pest resistance on the cotton plant. *Bt.* is a soil bacterium used extensively for pest control in organic agriculture. The MRC 7377 Bollgard 11 and MRC 7361 Bollgard 11 have higher yields and better quality cotton lint than the conventional Nigerian cotton which has a yield of 250–950 kg/ha. The adoption of Mahyco Bollgard II Hybrid seed has the potential to bring in much needed cotton revolution in the country. The released Mahyco Cotton Hybrids Bollgard II offers significantly higher yield levels, superior quality, and decrease in the cost of pest control, thereby enhancing the income level of farmers. Figure 7.2 shows a *Bt* cotton farm in NABDA, which was established for the confirmation of yield claims.

The revival of cotton farming will strengthen the economy of the country through employment generation, revival of related industry, and enhancement of lint exports. With an objective to meet at least part of the demand with local production by 2023–2025, the capacity building on seed production is ongoing in Nigeria. Scaling up is expected and this will include training registered farmers through various partners. By 2025, the objective is to create a significant pool of farmers who are adept in seed production protocols and at delivering the end product with 99% purity. The ultimate goal is for Nigeria to produce much needed *Bt.* Cotton seeds for local consumption and export.

7.2.1.2 Pod-Borer-Resistant (*Bt.*) Cowpea

Cowpea (*Vigna unguiculata*) is a food and animal feed crop grown in the semi-arid tropics covering Africa, Asia, Europe, the United States, and Central and South America. The grains contain 25% protein, and several vitamins and minerals. The plant tolerates drought, performs well in a wide variety of soils, and being a legume, it has the ability to fix atmospheric nitrogen. It is grown mainly by small-scale farmers in developing regions where it is often cultivated with other crops as it tolerates shade. It also grows and covers the ground quickly, preventing erosion. According to the Food and Agriculture Organization Corporate Statistical Database for 2016 (FAOSTAT, 2016), more than 5.4 million tons of dried cowpeas are produced worldwide, with Africa producing nearly 5.2 million. Nigeria, the largest producer and consumer, accounts for 61% of production in Africa and 58% worldwide. More than 11 million hectares are harvested worldwide, 97% of which are in Africa. Fifty-two percent of Africa's production is used for food, 13% as animal feed, 10% for seeds, 9% for other uses, and 16% is wasted (https://www.iita.org/cropsnew/cowpea/). Nigeria harvests 4.5 million hectares annually. The crop can be harvested in three stages: while the pods are young and green, mature and green, and dry. All parts of the cowpea crop are valuable because they are rich in nutrients and fiber for humans and livestock. In Africa, humans consume the young leaves, immature pods, immature seeds, and the mature dried seeds. The stems, leaves, and vines serve as animal feed and are often stored for use during the dry season.

The cowpea plant is attacked by pests during every stage of its life cycle. The cowpea aphid (*Aphis craccivora*) extracts juice from its leaves and stems at seedling stage and also spread the cowpea aphid-borne mosaic virus. Flower thrips (*Megalurothrips sjostedti*) attack the crop at flowering stage, thereby causing flower abortion; the pod sucking bugs (*Clavigralla tomentosicolis*) and pod borers (*Maruca vitrata*) attack the pods during post-flowering stage; and bruchids (*Callosobruchus maculatus*) attack the seeds during postharvest storage (Ugwu, 2020). The plants are also attacked by diseases caused by fungi, bacteria, and viruses. Parasitic weeds—*Striga gesneroides* and *Alectra vogelii*—choke the plants at all growth stages; while nematodes cause vascular occlusion, thereby preventing the absorption of nutrients and water from the soil. The cowpea pod borer poses the most serious constraint to cowpea production.

To address these challenges and provide further support to cowpea development in Nigeria, NABDA in collaboration with the African Agricultural Technology Foundation, Kenya, designed and constructed a level 2 containment facility at the Institute of Agricultural Research, Ahmadu Bello University, Zaria, where a cowpea transformation project was ongoing. The project resulted in the production of the pod-borer-resistant (PBR) cowpea also known as the *Bt.* Cowpea. The *Bt.* Cowpea has been field-tested in NABDA and IAR, as well as by Nigerian farmers to confirm the transformation. The *Bt.* Cowpea is currently awaiting commercial release. The newly improved *Bt.* Cowpea has the potential to resist Maruca attack; enhances farmers' ability to plant and harvest bountifully twice in a year with reduced chemical spray; and also reduces the cost of cowpea cultivation compared to the local variety.

7.2.2 Aeroponics Technology for the Production of Yam Seedlings

Aeroponics involves growing plants in the air or environment of mist. It does not require soil or an aggregate intermediate. Nutrient that is required for plant growth is transmitted through water. Operationally, the growing plant is suspended in a complete or partial enclosure and the dangling root is sprayed with nutrient-rich water. Vegetables such as lettuces and crops such as maize, potatoes, and yam could be grown using aeroponics (Mehandru et al., 2014).

Nigeria is the world's leading producer of yams, with an estimated production capacity of 18.3 million MT (in 1985) accounting for about 70%–76% of the world production and about 73.8% produced in Africa. By 2008, Nigeria's yam production doubled to about 35.02 million MT with value equivalent of US$5.654 billion. FAO (2018) estimated yam production for Nigeria at 49.53 million MT from a total land area of 5.99 million hectares (www.fao.org/faostat/en/#data/QC). One of the main challenges with yam production in Nigeria is seedling and seed yam production. To mitigate this challenge, International Institute of Tropical Agriculture (IITA) has undertaken numerous yam improvement projects with Nigerian researchers. Incubation programs have also been developed to transition identified high-yielding varieties from research to commercial level production. In 2020, IITA incubation program was reported to have generated a yield of about 23 tons/ha with a projected average yield of between 18 and 20 t/ha with good agronomic practices. This is a marked improvement from the average of 13 tons/ha that is prevalent in Nigeria (www.iita.org/news-item/iita-bip-team-sets-new-production-per-hectare-record-for-yam-farmers-in-nigeria).

NABDA is providing leadership in the deployment of aeroponics biotechnology for elite yam seed and seedling production. It is expected that this will improve the quality of the Nigerian yam with regard to uniformity of size and shelf life. The program is in collaboration with private sector (Biocrops Nigeria Limited). Already, hundreds of thousands of the elite yam seeds have been produced using this technology and distributed to farmers for trials in different locations and have also been planted in the BIODECs of NABDA, including the BIODEC in Ogbomosho. An ultra-modern facility is near completion in NABDA complex to enable annual production of five million (5,000,000) yam seedlings that will be distributed to farmers to meet the yam seed demand in the value chain. The objectives of the yam multiplication program using aeroponics include:

- Providing infrastructure for sustainable supply of yam seed and seedling to the Nigerian farmer;
- Increasing the production of consumer-preferred yam species to meet industry and market demands;
- Production of disease-free and high-yielding yam species that are exportable and with long shelf life; and
- Catalyzing of investment in the Nigerian yam value chain, upon which an estimated 31.8% of the country's population depend for food and income security.

With a projected increase of 30% yields and quality, the yam program will lead to the establishment of a sustainable, formal seed yam system with improved value of ware yam production, improving the livelihoods of at least 1.6 million people who depend directly on the yam value chain. A formal, traceable seed system also enables the export potential of Nigerian yams to the United States and Europe, which have high phytosanitary and quality standards. The overarching objective is to progressively commence large-scale production of yam seedlings by delivering 1,000,000 (1 million seeds/seedlings) in year one, and progressively move to achieve 100 million seeds/seedlings nationwide, taking advantage of the multiple locations of NABDA centers in 32 states of Nigeria.

7.2.3 Genomic Research and Innovation

Genomics is the underlying factor propelling modern biotechnology research and innovation. Full understanding of the genomes of the living organisms and the variation that exist between them is the basis its exploitation for agricultural, health, environmental, and industrial development. Nigeria has commenced work on studies that accelerates biotechnology to meet global demand for improving human and animal health and nutrition using genomic information. This is especially so in areas of national priorities including:

- **Food security:** For the first time in the world, Nigerian researchers at the National Biotechnology Development Agency-Center for Genomic Research and Innovation (NABDA-CGRI) have revealed the genetic basis for trypano-tolerance by the Muturu cattle breed (Abdulfatai et al., 2019). This has opened the door for the improvement of other breeds. With the full knowledge of the Nigerian cattle breeds, researchers are set to deploy genomic selection criteria to improve dairy and beef production from the Nigerian cattle.
- **Affordable medicines:** Genomics and Bioinformatics have been deployed for another look at the Nigerian plant biodiversity for cheaper and affordable medicines, beginning with anti-diabetics. NABDA-CGRI has elucidated the underlying biological mechanism of action and pushed out a Patent document for one very potent anti-diabetic candidate. In collaboration with partners, NABDA-CGRI is also working on re-purposing antiviral drugs from the Food and Drug Administration (FDA)-approved library for the treatment of Lassa fever.

7.2.4 Animal Genetic Resources

Nigeria is a signatory to Interlaken Declaration in 2007 that led to the preparation of Global Plan of Actions (GPAs) on Animal Genetic Resources (AnGR). The domestication of this plan of action was mandated on countries to mitigate genetic erosion, pollution, and dilution as currently witnessed in the livestock sector in Nigeria. This is to stem the tide of decreased productivity, breed extinction, sustainable loss of hybrid vigor, and low performance of animal agriculture and livestock value chain. The four major strategic pillars as recommended by Food and Agricultural Organization (FAO) for countries to sustainably manage their animal agriculture are:

a. Characterization, monitoring, inventory, and traceability;
b. Sustainable use and development;
c. Conservation; and
d. Policy frameworks, capacity building, and institutionalization.

Countries across the globe that adopted these strategies, especially Europe and North America, have been successful in managing their animal agricultural enterprises including global exports. In line with this, NABDA has been leading the process of the domestication of GPA since 2017 working with Federal Ministry of Agriculture and Rural Development (FMARD) and other stakeholders nationwide and abroad. Following a series of engagements and meetings based on comparative advantages and common national interest, NABDA and FMARD signed a Memorandum of Understanding (MoU) in 2018 to partner in bridging the gaps in National Animal Genetics. The collaboration was necessitated by the fact that the domestication and implementation of GPAs require advanced science especially biotechnological tools. Hence, the positioning of NABDA to provide the scientific support through research for development to FMARD in achieving the set global goals became desirable. The science-led process of this partnership recorded the following milestones:

I. Organization of National Stakeholders' engagement on Animal Genetic Resources sustainable management in 2017 with Ministries, Departments and Agencies (MDAs), Academia, Industry, Breeders Associations, Researchers, and other value chain actors;
II. Species-specific stakeholders' engagement on Bovine genetic resources with Cattle Breeders Associations including industry stakeholders like Friesland Campina (WAMCO) and SMAP Farm Ltd; acquisition of animals, equipment and consumables for livestock semen analysis; protocol development for cryopreservation of livestock semen; Creation of encrypted portal for National information and data management—www.angr.org.ng;
III. Facilitation of collaborations with FMARD on Semen Banking Facility, Mississippi State University, USA for Advanced Capacity building and international research, and SMAP Farm Limited for industrial and allied research;
IV. Establishment of First Animal Genetic Resources National Advisory Committee for the Country; and
V. Review and validation of National Animal Breeding Policy and preparation of National Strategy and Action Plan (NSAP) on Animal Genetic Resources in Nigeria.

AnGR research intervention using biotechnological tools is the panacea to fast, effective, and sustainable solutions in fixing the problems of livestock sector in Nigeria. Target areas include feed (Hydroponics), husbandry (Bull studs modeling), dairy (Milk product quality and safety assessment), beef (Meat quality and preservation for local market and exports), and exportation of genetic resources for forex. Hence, there is need for the following interventions in the sector:

i. Co-opting AnGR Research group into the National and State projects and programs with a focus on sustainable livestock development and transformation of the sector such as the national livestock transformation program implementation team of the National Council of State, Office of the Vice President, Economic Recovery and Growth Plan (ERGP) Team and FMARD;
ii. Establishment of National Animal Genetic Resources Bio-banking Facility to control the ongoing genetic dilution and pollution of livestock value chains;

iii. Special Purpose Vehicle (SPV) funding platform to support the acquisition of specialized equipment for the National AnGR Reference laboratory; and
iv. Validation of a national Clearing House for the issuance of genetic value assessment certificate for the use of local and imported animal genetic materials by breeders of Nigerian livestock species.

7.2.5 Hydroponics Fodder Production Technology

Hydroponics is a technology that enables plants to be grown in solutions containing nutrients that are utilizable by plants of interest. Hydroponic systems may or may not contain intermediate medium such as sand, sawdust, or gravel as sources of mechanical support. Hydroponics system approach to food crops and fodder production constitutes a sustainable solution offering opportunity to foster food and livestock feed security even in the face of climate change (Naik and Singh, 2013). As one of the viable options poised to deliver massive impact on food and feed production in Nigeria, hydroponics system guarantees a competitive return on investment for small and medium farmers. Hydroponics system for fodder production is an innovative approach integrating biotechnology as a tool to solve the problems associated with feed and food resources competition between herders and farmers in Nigeria. With a growing agitation for ranching, sustainability of feed resources in the ranches demands innovations such as use of grains for fodder production in hydroponics system.

The use of grains as quality feed in beef and dairy production is a common practice in the western world, where feed and food value chains are separated with little or no competitions. The potential of grains in improving the quality of livestock products in Nigerian livestock subsector is enormous for inclusion in ruminants feed. While the challenge remains the availability of grains for animal feed production, the deployment of hydroponics system for fodder biomass production provides a veritable tool for the exploration of less competitive grains. This will allow for the production of highly nutritious fodder biomass with enhanced feed quality in a resource efficient and effective ways. Some of the benefits of fodder from hydroponics to animals include:

- Faster weaning and maturity;
- Increased digestibility of fodder which leads to reduced greenhouse gas emission through feces;
- Enhanced immune system;
- Increased longevity and lifespan;
- Enhanced reproductive performances through early heat cycles;
- Improved fertility and fecundity; and
- Enhanced feed consumptions and appetite during heat stress.

NABDA is leading a national effort for the development of hydroponics with the development of prototypes for increased fodder production as ruminant feed using grains. Presently, three less competitive indigenous cereal crops, namely sorghum, millet, and wheat, are explored, with high preference for sorghum. This is partly

FIGURE 7.3 Cycle of hydroponics using sorghum gain to generate fodder for animal consumption: (a) Trays of fodder from germinated sorghum seed; (b) Different stages of sorghum seed development into fodder. Note: The day 1, sorghum seed immediately after soaking; days 2 and 3, seeds in the pre-germination (sprouting) period; days 4–6, germinated seeds.

because sorghum is a crop that is rich in antioxidants and high in fat. Hydroponically grown sorghum fodder has many nutritional advantages and, in fodder production, it is commonly used as a supplement to provide more fat (http://kaaa.co.ke/hydroponic-fodder-solution-farmers/). Using sorghum as grain, the hydroponics technology can be deployed for livestock fodder production at any scale and at different locations with a harvest period of only six days and grain to fodder conversion ratio of 1:12 (Figure 7.3).

Hydroponics technology for fodder production from seeds is an ongoing project in NABDA and the objectives are as follows:

- To promote sustainable all year-round animal agriculture and feed security with minimum environmental carbon footprints;
- To create a value chain that compensates for investment and stimulates investor's interest in livestock fodder biomass production with a 1:12 grain to fodder yield;
- Promote sustainable feed production in animal husbandry and contribute to the mitigation of free roaming of cattle on highways and other unauthorized premises across the country in the herders' bid to source for feeds;
- To support ranch development in all the different parts of the country on different scales of production; and
- To promote precision animal agriculture that will be business and investment friendly.

7.2.6 Aquaponics: Integration of Aquaculture and Hydroponics

Aquaponics is the combination of aquaculture (raising fish) and hydroponics (the soilless growing of plants) that grows fish and plants together in one integrated system (Lennard and Leonard, 2006). It allows for re-utilization of waste products across different farming systems (Tensingh et al., 2016). In line with its mandate to develop aquaculture sustainably and contribute to the food security of the nation, NABDA is currently carrying out several infrastructural development and research to infuse biotechnology and support entrepreneurship in aquaculture. Some of the researches

FIGURE 7.4 Aquaponics system installed at the NABDA headquarters: (a) and (e) are nurseries containing different stages of the plant; (b) and (c) Re-circulatory aquaponics system containing different stages of fish; (d) Gravel bed ready for the transplant of plants in (a) and (e); tank in (d) holds wastewater from (b) and (c), which is filtered through the beds in shown in (d), (a), and (e).

are being undertaken in collaboration with the Institute of Aquaculture, University of Stirling, Scotland, UK. The collaboration has aided NABDA in the development of 13 polymorphic microsatellite markers for genetic diversity studies in a population of *Clarias gariepinus*, a common fish species in Nigeria. The partnership has also enable NABDA to detected24 SNPs (single nucleotide polymorphisms) discriminating *Clarias gairepinus* from *Clarias anguillaris* the first ever-diagnostic molecular markers for the fish species in the globe.

NABDA has great interest in fish nutrition (fish feed) research and development, and has conducted trainings for Nigerians on proximate analysis and fish health. In this respect, NABDA has acquired equipment for proximate analysis of raw materials and feeds for research and development, and for commercial use. In addition, the Agency is developing the integration of aquaculture and hydroponics to further enhance the sustainability of fish farming by reusing fish wastewater in production of vegetables such as tomatoes, cabbage, and lettuces as shown in Figure 7.4a–e.

7.2.7 Bioethanol Development Programme

Nigeria has very few functional bioethanol plants and much fewer at industrial scale. One of the newest bioethanol facilities in Nigeria is the Allied Atlantic Distilleries Limited (AADL), Igbesa in Ogun State, which produces about 9 million L/annum. According to National Bureau of Statistics, Nigeria consumed 54.3 million L of

Premium Motor Spirit (PMS) daily within the first quarter of 2017. This requires 5,430,000L of ethanol per day to make E10 gasohol (90% gasoline with 10% ethanol), amounting to 1,981,950,000L of ethanol per year. This can be supplied by converting wastes from 40% of the annual 57 million ton of cassava to ethanol. Currently, Nigerian local refineries are producing only about 8 million L of PMS per day. This requires only 800,000L of ethanol per day (292,000,000L per annum) and can be supplied by converting wastes from only 5.84% of cassava produced in Nigeria (www.irena.org).

This makes cassava an important resource for strengthening Nigeria's capacity for ethanol production is cassava.

While several efforts have been directed at increasing cassava production, the current production capacity in the country is limited and hardly meets the feedstock requirement. This is partly due to demand for cassava as food: cassava is the source for garri, which is one of the staple foods in Nigeria. There is increasing efforts to diversify feedstock for biofuels production to reduce the competition between food and raw materials. Sugarcane on the other hand has a higher prospect for use as ethanol feedstock due to the fact that relatively more states have the appropriate climatic conditions for its cultivation. Hence, yield is expected to be high within fusion of technology and inputs.

Considering the production gap in the Nigerian industrial ethanol sector (400 million L) and the millions of liters/annum required for meeting the E10 policy, there is increasing need for the establishment of small- to medium-scale production facilities of bioethanol in Nigeria. Hence, NABDA is leading national efforts aimed at developing and adapting the ethanol technology to promote entrepreneurship and startup. An example of this effort is the establishment of 1,000L per day pilot ethanol plant in Ogbomosho (Figure 7.5) that uses cassava for the production of ethanol. The following are the specific goals and objectives of the project.

- Reducing Nigeria's carbon foot print and maximize carbon credit opportunities;
- Producing alternative fuel that is environmentally friendly;
- Creating rural wealth and jobs;
- Establishing a thriving home-grown industry; and
- Reducing Nigeria's dependence on importation to meet ethanol demand.

FIGURE 7.5 A 1,000 L/day pilot ethanol plant at the NABDA Bioresources Development Centre in Ogbomosho, Oyo State: (a) Distillation chamber/column; (b) Fermentation chambers.

7.2.8 Bioenergy Development Programme

Anaerobic Digestion (AD) is a biotechnique that enables the production of vital compounds from biomass. The production of methane from biomass using AD technology is called biomethanization. Biomethanization produces biogas, which principally comprises methane (CH_4) and carbon (IV) oxide (CO_2). Biogas may contain hydrogen sulfide; traces of gases such as ammonia and chlorinated compounds; and digestate that is rich in nutrients such as phosphate, nitrogen, and trace elements and could be used for soil enhancement (Verma, 2002). AD technology has universal acceptance in biomass resources re-utilization for biomethane and biofertilizer production (Onyia et al., 2001). However, biomethanization needs regional adaptation to meet unique substrate characteristics, purification of the biogas to enrich the biomethane production and determination of the biomethane potentials of local biomass feedstocks to enable co-digestion formular development.

The Bioenergy Development Programme (BDP) was designed to adapt biomethanization technology to the regional biomass peculiarities to enable deployment of the technology for waste management, bioenergy, and biofertilizer production. The BDP seeks to meet Nigeria's target of 5% biomass renewable energy contribution to energy mix. The program embodies components designed to ensure that the broader objectives of the generation and deployment of biomass energy in power generation, waste management, and rural domestic heating are sustainably met. The BDP has some major components and these are discussed below.

7.2.8.1 Design and Development of Prototype AD Digesters

Prototype AD digesters have been designed and developed in Nigeria for use in households, restaurants, markets, and small- and medium-scale enterprises that generate biomass resources as by-products. Figure 7.6 shows the three different sizes of the digesters that have been designed and produced by NABDA scientists

FIGURE 7.6 Prototype biodigesters produced in Nigeria for the production of bioenergy used in power generation and domestic heating (credit: NABDA). The digesters where produced by scientists and engineers at NABDA and have been patented. The 250L and 500L digesters are to serve households, restaurants and small farms; while the 1,000-L digester is intended to produce electricity and heating in markets and large farms.

and engineers. The feat was achieved after due consultation with stakeholders including government MDAs; international organizations such as United Nations Industrial Development Programme (UNIDO), and NGOs such as the Waste Management Society of Nigeria (WAMASON). The product has been patented (NG/P/2019/187) and is ready for commercial deployment. The unique features of the product include:

- Digester body and other components such as scrubber, reinjection system, and the discharge system are made of polypropylene to enable the facility withstand high ambient temperature and to confer durability;
- Integrated scrubber for a purer biomethane through the removal of hydrogen sulfide and carbon dioxide;
- Pneumatic stirring system to reduce the need for moving components within the digester and enhance ease of use without compromising process efficiency;
- Integrated hydrolysis tank to enable pretreatment of biomass prior to feeding to the main digestion tank;
- Effluent discharge point for ease of digestate removal during continuous feeding operation; and
- Pressure monitor and gas flare system.

7.2.8.2 Design and Development of Biomethane Test Systems

A critical step toward efficient development of Nigeria's biomass resources for electricity generation and heat production is the determination of the methane potential of biomass feedstock used for biomethane production. Nigeria has abundant biomass resources that could be used for methane production. However, the efficient use of these bioenergy resources demands knowledge of their methane content. Such knowledge enables the combination of substrate to compensate for biomass characteristics that might be lacking when single substrates are used in biomethane production.

Biomass characterization can easily be carried out by determining the carbon, nitrogen, and phosphorus contents. More advanced characterization could further determine metal content for the purpose of enhancing the biomethane generation process through biocatalysis. However, in Nigeria, the critical challenge has been to determine the methane potential of biomass. This challenge is now being met through concerted effort at NABDA. Figure 7.7 shows biomethane test system developed in Nigeria for the determination of the methane potential of substrates (also known Gas Potential 21 or GP-21), because it operates for 21 days as a standard operating duration.

The test system will enable the development of a biomass co-digestion matrix and enhance the efficiency of the combination of biomass resources for bioenergy generation. It will also enable researchers to map the best- and worst-scenario bioenergy potential of the country based on methane-yielding potential of the abundant biomass resources. Finally, it empowers the country to make realistic projections on what proportion of biomass energy could be integrated into the energy mix based on known biomass abundance and methane yield potentials (Ezebuiro and Koerner, 2017).

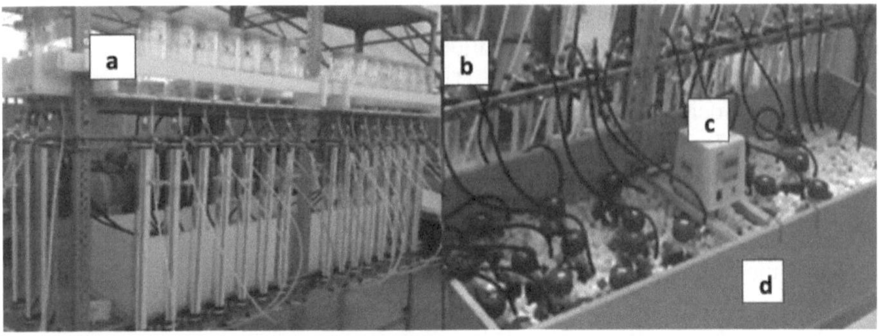

FIGURE 7.7 Biomethane test system for the determination of the methane potential of biomass feedstock (Credit: NABDA): (a) Set of eudiometers containing barrier solution through which biogas is collected; (b) Set of 1 L reactors that are connected to the eudiometers (a) through tubes; (c) Thermostat to keep the water bath at constant temperature; and (d) Water bath containing water and 1 L reactors at experimental temperature.

7.2.8.3 Adaptation of Existing Power Generating Systems to Utilize Biomethane

In addition to the two main achievements of the BDP of Nigeria, human capacity building in energy systems retrofitting to utilize biogas or biomethane has been extensively carried out. Retrofitting of the carburetor of petrol utilizing generator sets to use biomethane is a key activity of the BDP. This is expected to aid seamless transitioning to biomethane fuel. This is particularly important given that the program also aims to deepen the penetration of biomass energy in rural electricity generation, and heating to reduce indoor pollution and felling of trees.

7.2.8.4 Development of a Commercial Model for Bioenergy Systems Deployment

Considering that Nigeria has patented (NG/P/2019/187) its bioenergy digesters (Figure 7.7), large-scale production and deployment become imperative to achieve the set target for bioenergy proportion in the national energy mix. In view of this, scientists, engineers, and business analysts at NABDA, UNIDO and WAMASON are working together to develop an entrepreneurship model for the digesters that will be market-driven. Some of the likely considerations to keep the cost low and affordable are linking the program to the global emission reduction program through the carbon credit scheme and leveraging the tax and import duty incentives of the Federal Government for renewable energy entrepreneurs.

7.2.9 National Microbial Culture Collections Programme

Microbial Culture Collections (MCCs) are Biological Resource Centres (BRCs) necessary for the advancement of biotechnology, sustainable development, and promotion of bio-economy. They serve as repositories for sustainable conservation and utilization of microbial genetic resources and as support systems for biotechnology innovations in health, environment, bio-industries, agricultural systems, and renewable energy. Besides keeping type strains and conducting research, MCCs maintain the database and information on microbial genetic materials and isolated strains (Overmann, 2015). During several national stakeholders' consultative fora, sensitization workshops, seminars, and dialogues, NABDA, as the National Designate Authority on Biotechnology, was chosen as the most appropriate Agency of Government to initiate and host an MCC program for the country. This necessitated the commencement of the NMCC program by NABDA and its inclusion in the national budget in 2013.

7.2.9.1 Roadmap and Strategic Objectives of the MCC Program

Some of the critical Resolutions of the National Stakeholders' Workshops and Dialogues informed the Roadmap and strategic objectives which include:

i. Strict compliance with existing standards, regulations, and methods of establishing MCC and collecting microbes;
ii. Establishment and management of MCC in the South-South region (since the University of Port-Harcourt donated land for the conservation of microbial resources implicated in bioremediation of the highly polluted Niger Delta region) and other identified Institutions in the Country with a Central Core Facility located in NABDA Headquarters where the national database would also be domiciled;
iii. Regular National Stakeholders' Sensitization Workshops, Experts' Meetings, Roundtables, and Regional Dialogues to ensure periodic updates, evidence-based and inclusive decisions;
iv. Extensive human capacity development to build the critical mass of professionals required for successful implementation of the program; and
v. NMCC web portal and database development for online archiving of sequenced indigenous microbial strains among others.

7.2.9.2 Achievements of the MCC Program

Whereas completing the MCC facility in Port-Harcourt has experienced setbacks due to technical design issues and terrain-related challenges, significant progress has been made on the other objectives of the program. Some of these achievements include:

a. Development of a web portal for Nigerian MCCs program (www.nmcc.com.ng);
b. Creation of indigenous sequenced MCCs Information Database;
c. Capacity building of staff on emerging techniques in culture collections, characterization, identification, authentication, and preservation;
d. Hosting of National Experts` Meeting and Roundtable, and South-West Regional Dialogue on the Status and strategies for a successful implementation of the MCC program in 2018 and 2019, respectively;
e. Collaboration and linkage with the Centre for Agriculture and Biosciences International (CABI), United Kingdom and relevant national institutions; and

f. Skeletal fungal collection, identification, and preservation at the temporary MCC back-up laboratory at the Environmental Biotechnology and Bio-conservation Department at the NABDA headquarters to enhance effective utilization of available equipment, facilities, and human resources.

7.3 APPLICATION OF RECOMBINANT DNA TECHNOLOGY

Recombinant DNA is the term applied to chimeric DNA molecules that are constructed in vitro, then propagated in a host cell or organism. The basic recombinant DNA consists of a vector and an insert. The vector is a replicon capable of replicating in the cells of choice. It is endowed with a functional replication origin. The replication origin usually carries a selectable marker, and is typically engineered to conveniently accommodate inserts. Vectors are based on naturally occurring replicons, such as bacterial plasmids, viruses, or cellular chromosomes. Inserts can be of any sort—long or short segments of DNA, from natural or synthetized sources. The resulting recombinant DNAs are often referred to as recombinant clones, which is shorthand for chimeric DNAs that are isolated in cellular or viral clones; and the process of producing these recombinants is frequently called DNA cloning or gene cloning (Phuc, 2018).

7.3.1 DNA Typing

NABDA, National Institute of Science laboratory Technology (NISLT), and National Institute of Pharmaceutical Research (NIPRID) are collaborating on exploiting the DNA typing tool (technology) for wider applications and for revolutionizing developmental programs in health care, immigration, prisons, judiciary, defense and security, and crime detection and prosecution. Advanced equipment for conventional DNA laboratory and facility that will possibly serve as a reference point for DNA typing in Nigeria has been established in Nigeria. The Forensic Science facility will support various sectors including genetic disease investigation, judiciary for paternity determination, and emergency management agencies for fire and accident victim's identification. Other applications include security for identification of criminals and crime suspects as well as database for Prisons and Immigration Services.

The need for Certified Forensic Reference Laboratory and DNA database and management is high. The existence of such facility will enhance the country's acceptance in the global fight for crime, immigration, and border control as well as integration into the global networks such as International Criminal Police Organization (INTERPOL), World Anti-Doping Agency (WADA), and United Nations Office on Drugs and Crime (UNODC). Nigeria currently does not have any established criminal forensic interoperable database, a necessary requirement for the collaborative fight against local and international crime. Considering these deficiencies and the nature of the program, multi-stakeholder meetings have been held multiple times, with the relevant parties assuming various responsibilities including:

- Building the highest capacity in toxicology, drugs and precursors, firearms, shoe marks, and fingerprint identification, and other physically based analysis;

- Developing highest standards in forensic DNA, database hosting, and serving as reference laboratory;
- Coordinating forensic practice and research including requirements for the establishment of forensic laboratories;
- Database management and Forensic Science legislation for proper legal backing of technology deployment;
- Facilitating private sector involvement in building national and African genomic database. Currently, a biotech startup named 54gene is setting the pace for improved private sector participation in bio-banking;
- Building of requisite skills that can be deployed to support other relevant agencies or institutions such as Nigeria Police Force, Nigerian Army, Immigration, Ministry of Youths and Sports, Nigerian Prisons Service, Ministry of Justice, Nigerian Drug Law Enforcement Agency (NDLEA), National Population Commission (NPC), Civil Defense, Nigerian Medical Association (NMA); and
- Facilitating stakeholders' workshop which would serve as a forum for the development of the draft bill and constituting a working group in partnership with the Molecular Diagnostics and DNA Forensics Working Group of the BSN. This workshop has been organized and had the following outcomes:
- Development of Forensic Science draft bill;
- Development of Coordination Framework for National Forensic DNA Database and Management (Certification, Accreditation of Forensic Science personnel, Skillset requirement and training, Equipment and Standard Operating Procedures (SOPs));
- Streamlining funding mechanism; and
- Sensitization and awareness creation for Forensic Science in Nigeria.

7.3.2 Development of Level 3 Biosafety Laboratory and Bio-Productions

Biosafety level 3 facility is appropriate for work involving microbes, which can cause serious and potentially lethal disease via inhalation. This type of work can be done in clinical, diagnostic, teaching, research, or production facilities. The precautions undertaken in BSL-1 and BSL-2 labs are followed, as well as additionally stipulated international measures (U.S. Department of Health and Human Services, 2009):

- All laboratory personnel are provided medical surveillance and offered relevant immunizations (where available) to reduce the risk of an accidental or unnoticed infection;
- All procedures involving infectious material must be done within a biological safety cabinet with HEPA (High-Efficiency Particulate Air) filter;
- Laboratory personnel must wear solid-front protective clothing (PPE gowns that tie in the back, powder-free nitrile gloves, etc.). These cannot be worn outside of the laboratory and must be discarded or decontaminated after each use; and
- A standard incinerating facility must be provided, well distanced from the laboratories and office for waste disposal and destruction of contaminated items.

Biosafety level 3 facilities are commonly used for research and diagnostic work involving various microbes, which can be transmitted by aerosols and/or cause severe disease. These include *Francisella tularensis, Mycobacterium tuberculosis, Chlamydia psittaci, Venezuelan equine encephalitisvirus, Eastern equine encephalitis virus*, SARS coronavirus, SARS-Cov-2, MERS coronavirus, Rift Valley fever virus, *Rickettsia rickettsii*, several species of *Brucella, chikungunya*, yellow fever virus, West Nile virus, and *Yersinia pest* (U.S. Department of Health and Human Services, 2009).

To respond to disease outbreaks from some of these causative agents, research into and development of vaccines are required. Vaccine development is a multi-stakeholder activity requiring participation of scientific experts and regulators. Hence, the Nigerian Immunology and Infection Network has been established in Nigeria to respond to the control of infectious pathogens, emerging diseases, and training needs. Biosafety level 3 facilities have been installed in different institutions including NABDA. Activities of the Biosafety Laboratory Programmes and the Nigerian Immunology and Infection Network will enable coordinated research for the development of candidate vaccines, containment of pathogens and development of rapid-point-of-care diagnostic test kits. It will also support the efforts of Ministries of Health (disease control) and Defense (biosecurity).

The establishment of the biosafety/biosecurity laboratory is in line with the aim and programs of the National Action Plan for Health Security (NAPHS) launched in 2018. NABDA is actively working with the National Center for Disease Control (NCDC) and Ministry of Defense on the action plan. Deploying technologies for recombinant vaccines and molecular diagnostic kits production requires high level facility, coordination and international expert skills transfer, which is why networks of experts are required. Such networks include the immunology network and the Centre for Science and Technology of the Non-Aligned and Other Developing Countries (NAM S&T Centre), to which countries such as Egypt, India, Myanmar, Nepal, Nigeria, Palestine, Peru, Sri-Lanka, and Vietnam are members. In Nigeria, NABDA is the focal point for the NAMS&T program and anchors the NABDA Expert Fellowship Programme. NABDA is also hosting the Nigerian Network of Immunology for networking, immunology/vaccine experts' database; training of next-generation vaccinologists. NABDA also provides scientific background for first responders in disease outbreak situation, especially those of immune-related.

7.3.3 Production of RNA Extraction Kit for SARS-COV-2 Identification

Only recently, Nigeria successfully validated locally produced RNA extraction kit for the identification of SARS-Cov-2, the causative agent for COVID-19 (Gorbalenya et al., 2020). Validation of the Ribonucleic Acid Extraction Test Kit would enable Nigeria to locally produce cheap testing kits for COVID-19. The project is a partnership involving NABDA and NCDC. NABDA is providing the national leadership for the project while validation activities are carried out by NCDC; and the University of Sheffield serves as the technical partner. The test kits will enable Nigeria and other African countries to significantly scale up citizens' testing for COVID-19, especially the working population in critical sectors such as agriculture and health.

7.4 CONCLUSION

Nigeria is obviously shifting from conventional technologies to modern biotechnology in areas of food security, healthcare service provision, environmental stewardship, and industrial production. Agricultural sector has received the most adaptation and deployment of biotechnology in Nigeria, mainly in the form of introduction of *Bt.* crops and animal genetic resources management. Few applications, especially recombinant DNA technology, have been deployed in the health sector for vaccines and diagnostics production. Efforts to bring institutions together to harness expertise are ongoing; and the role of the BSN in galvanizing and providing networking among professionals involved in the different aspects of biotechnology is pivotal. There is a high level of assurance that many more biotechniques will be mainstreamed into daily national life, especially as it concerns services such as citizens' identification in paternity testing, etc. There is limited application of biotechnology in advanced healthcare service delivery such as stem cell therapy or tissue cloning yet in Nigeria. This underscores the need for increase in human capacity and infrastructure. Finally, increased collaboration with private sector is providing the opportunity to expand biotechnology applications as a panacea for industrial revolution and national development.

REFERENCES

Abdulfatai, T., Utsunomiya, Y., Tani, E. G., Nashiru, O. and Hanotte, O. (2019). Genome sequence analysis reveals selection signatures in endangered trypanotolerant West African muturu cattle. *Frontiers in Genetics* 10:442, DOI:10.3389/fgene.2019.00442.

Ezebuiro, N. C. and Koerner, I. (2017). Characterization of anaerobic digestion substrates regarding trace elements and determination of the influence of trace elements on the hydrolysis and acidification phases during the methanization of a maize silage-based feedstock. *Journal of Environmental Chemical Engineering* 5(1):341–351.

Fadare, T. A. and Amusa, N. A. (2003). Comparative efficacy of microbial and chemical insecticides on four major lepidopterous pests of cotton and their (insect) natural enemies. *African Journal of Biotechnology* 2(11):425–428. DOI:10.5897/AJB2003.000-1085.

Gorbalenya, A. E. Baker, S. C., Baric, R. S., de Groot, R. J., Drosten, C., Gulyaeva, A. A., et al. (2020). The species severe acute respiratory syndrome-related coronavirus: classifying 2019-nCoV and naming it SARS-CoV-2. *Nature Microbiology* 5(4):536–544. DOI:10.1038/s41564-020-0695-z. PMC 7095448. PMID 32123347.

Lennard, W. A. and Leonard, B. V. (2006). A comparison of three different hydroponic subsystems (gravel bed, floating and nutrient film technique) in an aquaponic test system. *Aquaculture International* 14(6):539–550. DOI:10.1007/s10499-006-9053-2.

Mehandru, P., Shekhawat, N. S., Rai, M. K., Kataria, V., Gehlot, H. S. (2014). Evaluation of aeroponics for clonal propagation of *Caralluma edulis*, *Leptadenia reticulata* and *Tylophora indica* – three threatened medicinal Asclepiads. *Physiology and Molecular Biology of Plants* 20(3):365–373.

Moseley, W. G. and Gray, L. C. (2008). *Hanging by a Thread: Cotton, Globalization and Poverty in Africa*. Ohio University Press and Nordic Africa Press. ISBN 978-0-89680-260-5.

Naik, P. K. and Singh, N. P. (2013). Hydroponics fodder production: an alternative technology for sustainable livestock production against impeding climate change. MTC on Management Strategies for Sustainable Livestock Production against Impending Climate Changes. pp. 70–75.

National Biotechnology Policy (2001). *Federal Ministry of Science and Technology, Nigeria.*

Onyia, C. O., Uyub, A. M., Akunna, J. C., Norulaini, N. A., and Omar, A. K. M. (2001). Increasing the fertilizer value of palm oil mill sludge: bioaugmentation in nitrification. *Water Science and Technology* 44(10):157–162.

Overmann, J. (2015). Significance and future role of microbial resource centers. *Systematic and Applied Microbiology* 38:258–265.

Phuc, V. P. (2018). Chapter 19. Medical biotechnology techniques and applications. In: Debmalya, B. and Vasco, A. (Eds.) *Omics Technologies and Bio-Engineering,* pp. 449–469. London: Academic Press. DOI:10.1016/b978-0-12-804659-3.00019-1

Stewart, N. C. (2016). *Plant Biotechnology and Genetics: Principles, Techniques, and Applications.* John Wiley & Sons, pp. 227–228. ISBN: 1118820126, 9781118820124.

Tensingh, G. P., Rachel, J. E. and Muthuramalingam, T. (2016). *Low Cost Hydroponic Device. New Technologies for Livestock Farmers - Introduction and Release.* University Research Farm, Tamil Nadu Veterinary and Animal Sciences University, pp. 28–30.

U.S. Department of Health and Human Services (2009). *Biosafety in Microbiological and Biomedical Laboratories*, 5th ed., pp. 30–59; 123–289. HHS Publication No. (CDC) 21-1112. Health and Human Services: USA.

Ugwu, J. (2020). Insecticidal activity of some botanical extracts against legume flower thrips and legume pod borer on cowpea *Vigna unguiculata* L. Walp. *The Journal of Basic and Applied Zoology* 81(1):13. DOI:10.1186/s41936-020-00153-3.

Verma, S. (2002). *Anaerobic Digestion of Biodegradable Organics in Municipal Solid Wastes.* New York: Columbia University. pp. 10–15.

WEBSITES

1. http://www.cottontapafrica.org/nigeria.html. Accessed 25/5/2020.
2. www.fao.org/faostat/en/#data/QC. Accessed 5/6/2020.
3. http://kaaa.co.ke/hydroponic-fodder-solution-farmers/. Accessed 5/27/2020.
4. https://www.iita.org/cropsnew/cowpea/. Accessed 5/27/2020.
5. https://www.iita.org/news-item/iita-bip-team-sets-new-production-per-hectare-record-for-yam-farmers-in-nigeria/. Accessed 5/27/2020.
6. Food and Agriculture Organization of the United Nations Statistics Division [FAOSTAT] (2016). Available at: http://faostat3.fao.org/download/Q/QC/E. Accessed 7/2/2021.
7. https://www.irena.org/-/media/Files/IRENA/Agency/Events/ 2018/Jan/Sustainable-bioenergy/213-Dr-Ogbonna-et-al-Nigeria-U-JIRCAS-Ethanol-from-Cassava.

8 Communicating Scientific Innovations
The Case of Agricultural Biotechnology in Nigeria

R.S.M. Gidado, A. Isah, and S.O.O. Iweajunwa
National Biotechnology Development Agency (NABDA)

N. Muchiri
African Agricultural Technology Foundation

CONTENTS

8.1 INTRODUCTION

The integration of scientific innovation and its application solves basic issues that have confronted major aspects of human life ranging from medicine, food, textile to the environment (Aina et al., 2019), which has resulted in an urgency in expanding the scope of science communication (Jonathan, 2017). The integration of scientific information with personal values is expedient in enabling people take the right steps when making important life decisions such as vaccinating their children,

DOI: 10.1201/9781003179177-8

ascertaining the safety of foods (including the application of modern biotechnology in agriculture), how to reduce the negative impacts of climate change and to contribute to national developments (Jonathan, 2017). The act of effective science communication (SciCom), however, is a multifaceted task that requires the acquisition of skills (The National Academies of Sciences, Engineering and Medicine, 2017). Among the several scientific tools and processes, genetic engineering, a major tool in modern biotechnology, remains one of the most debated technologies (Jennifer, 2016; Arthur, 2003), while the alteration of plants' hereditary properties is among the most contentious branch of bioengineering (Sigrid, 2013).

The intense continuous debates surrounding the adoption and acceptance of agricultural biotechnology in Nigeria and the world at large have placed focus on other variables beyond science itself (Mariechel, 2003). Public opinion about science, particularly agricultural biotechnology, has continued to be affected by the polarization of stakeholders which is normally caused by divergent information sources and conflicting and mixed messages. Findings by Navarro (2011) have highlighted communication as one of the several key variables required for the creation of an enabling environment for agricultural biotechnology.

The need for a more rapid and holistic science communication in Nigeria has never become more necessary than now: with an estimated 263 million population by 2030 (United Nations, 2015), increasing rate of hidden hunger (Uche and Familusi, 2018), pressure on the government to meet the Sustainable Development Goal (SDG) of zero hunger by 2030 (Gil et al., 2019), emerging disease pandemics/epidemics/endemics, and the continuous spread of scientific myths (Letrud and Hernes, 2019), science communication that is centred on public enlightenment has become an emergency in Nigeria. Survey conducted by the Open Forum on Agricultural Biotechnology (OFAB) in Africa, Nigeria Chapter (2019) has underscored 'ignorance' as the biggest threat to the advancement agricultural biotechnology in Nigeria. Mutual understanding and dialogue were highlighted by the International Service for the Acquisition of Agri-biotech Applications (2020) as part of the key factors that will enable the global community to understand the attributes of agricultural biotechnology and ensure public acceptance and adoption.

SciCom is the production of opinion-forming, sensitization and public understanding through the engagement of a well-patterned skill such dialoguing and targeted public engagement that is driven towards public acceptance and appreciation (Burns et al., 2003). Several scientific innovations and inventions that are now part of our everyday life have cut across the global stage to the very remote settlements around the world. Food security, increased human life span, enhancement of work-play balance and improved life quality are some of the benefits that have accrued from scientific discoveries and processes (Navarro et al., 2013). Ironically, such discoveries underwent different degrees of scepticism, resistance, uncertainty and fear before they were eventually accepted.

Over the years, the scientific inventions and technological revolution have always galvanized undivided attention from policy makers and the public in general (Einsiedel, 2008). Scientific findings by Walker (2011) have underscored 'trust' as a key factor that determines the acceptance of any new technology or scientific discoveries even when such technology or discovery poses no harm to man and his environment. Trust is a key factor because the perception of risks differs not only between scientists and laymen but also among countries and cultures. Adequate science communication is a fundamental practice that must be done if the trust of stakeholders and the general public must be secured for any new scientific discovery and innovation.

Introducing topics that centres on novel scientific innovation is a natural way of triggering dialogue during public engagement activities (Devos et al., 2007). The sceptical and indecisive attitude of some Nigerians towards food produced using agricultural biotechnology tools is a strong indication that assessment of products derived from agricultural biotechnology needs to move from scrutiny and risk-centred policy to a more fact robust analysis. Research findings by Hallman (2008) has also concluded that adequate science communication that will unveil how new scientific discoveries and technological innovations have shared societal vision is very necessary.

Key components of science communication have been highlighted by Irwin (2009) to include trust that is earned as a result of transparency. Arntzen et al. (2003) have also highlighted that it is not enough that biotechnology directly benefits the lives of the public, but its credibility and trust by the public and stakeholders including the government through science communication and transparency is necessary. Holliman et al. (2009) have underscored the need for the enlargement of the scope of science communication to go beyond public understanding to public engagement with science, stating that, it is not enough to think that the ignorance by the public on science-related matter could be resolved by mere provision of scientific information, rather to make sure that the public have an active role in science by making sure that their voices are heard. The urgent need to, therefore, deploy science communication to increase societal trust on emerging scientific innovations and theories is vital if the United Nations SDG 2017 of Zero hunger by 2030 must be met in Nigeria.

8.2 THE DIVERSITY OF SCIENCE COMMUNICATORS

The communication of scientific information to the public can be carried out by individual scientists including health professionals and other researchers, agencies of the federal government that have been mandated to promote the safe use of scientific innovations such as the National Biotechnology Development Agency (NABDA), trained media practitioners such as science journalists, advocacy organizations and corporate organizations that are research oriented. Individuals who are science enthusiastic and advocates can also be an active science communicator.

It is the right of the public to know and understand the entirety of new scientific innovations and to relate their concerns and reservations to the pioneering scientists. There is a need for research scientists to be outspoken on the safety, benefits and general applications of their research findings in order to facilitate the quality of public discourse and unravel the relevance of science to society.

8.3 ETHICAL CONSIDERATIONS IN SCIENCE COMMUNICATION

Ethical consideration, one of the critical components of science communication, involves the choice of timing, appropriate words, manner of approach, dress style and voice intonation to be used that will best suit the audience or individual in view. For the communication to flow steadily, the scientist or communicator must also be a good listener in such a way that he is able to pay undivided attention to his audience. Ability to pay attention to the facial expression of the audience or individual is also very key for a successful communication of science. Bottom of Form

8.3.1 Public Engagement

The guidelines on the broad concept of public engagement that was underscored by Leshner (2003) have revealed that science communicators need to be transparent in terms of scientific product display and application in order to engage the public in a bidirectional manner. Transparency during science communication entails that both the benefits and side effects, if any, of such scientific innovations should be revealed to the public, neither the negative nor positive side of such product should be over-stressed. The targeted audience should be given the choice of making their decision without any form of lobbying or enforcement. Though, it may not be possible to get to convince the whole audience at a go, but the convinced ones among the audience should be trained to train the others by engaging them further on such scientific innovations in the future or at any given opportunity. The communicator must engage the public in such a way that their concerns and perspectives are adequately taken care of by providing the public with adequate responses such as giving them information materials, granting them access to website links where they can go and do further study on their own, inviting them to the research institutes to see some of the basic process involved in the production of such scientific product if possible, and releasing some of the products to them for them to use and experience the benefits themselves like in the case of *Bt Cowpea* in Nigeria where some selected farmers were given the seeds to plant in order for them to have a first-hand experience with the crop (Leshner, 2003).

Significant results are usually achieved whenever scientists participate in group or public discussions sharing their research experiences and addressing concerns raised by the audience in such a manner the choice of words used are familiar to the hearers (Rowe and Frewer, 2004).

However, as engagement or dialogue with stakeholders continued to gain importance due to the interface between science, politics and public uncertainty, researchers, including Stilgoe and Wilsdon (2009), have reiterated the necessity for the engagement of the upstream that will facilitate a productive and upbeat discussion. Though agricultural biotechnology has been identified by Navaro (2011) as one defining technology that has transformed the interplay sandwiched between science and society, the application of modern biotechnology tools in agriculture has continued to undergo thorough scrutiny by the public among other innovative technologies (Michael et al., 2002). Public debate about agricultural biotechnology has continued to spring (Navaro, 2011) – notwithstanding the fact that on the 23rd year of commercialization of biotech/GM crops in 2018, 26 countries grew 191.7 million hectares of biotech crops worldwide (of which over 90% are from developing countries) (ISAAA, 2018). As a result of the strong positive, negative and indifferent responses of the different publics to agricultural biotechnology, despite its tremendous benefits to mankind, Castillo (2003) has concluded that a strong correlation exists between public opinion and the acceptance and adoption of any innovative technology such as agricultural biotechnology. The right to know and to also participate in every science-related decision in Nigeria has been emphasized by stakeholders, since such scientific decisions may have direct or indirect impacts on their wellbeing.

The spread of fear by the opponents of agricultural biotechnology in Nigeria has caused a lot of panic among the public. For instance, Poortinga and Pidgeon (2007) and Peters and Sawicka (2007) have cited the way shops in Europe were threatened by some consumers and non-governmental organizations to remove products of agricultural biotechnology (particularly Genetically Modified Organisms (GMOs)), from shelves in supermarkets. Similar occurrences were also recorded in India, Thailand and Philippines, where policy makers were influenced by civil society groups to impose a moratorium on GM research and terminate the release of a biotech crop despite regulatory approval. The only way such fears can be stopped is to create a platform where the public can have access to accurate information and proper enlightenment on the technology. In Nigeria for instance, the OFAB in Africa, which is an initiative of the African Agricultural Technology Foundation (AATF), was created to stand as a platform for the dissemination of the massive and tremendous information on the benefits, safety and applications of agricultural biotechnology and products (GMOs) to important stakeholders of the agriculture and food sectors such as the various agricultural ministers and commissioners, scientists, policy makers, regulators of the technology, journalists, civil society groups, industrialists, legislators, academics and the direct consumers (AATF, 2017). These various stakeholders are brought together on the platform of OFAB Nigeria for the exchange of information on agricultural biotechnology, where they are all able to share their knowledge, concerns, interests and experience (AATF, 2017). Communication is the leading factor among the four major factors identified by Ezezika et al. (2012) that appears to influence agricultural biotechnology in Sub-Saharan Africa including Nigeria, the other three being commercialization, culture and religion.

The Food and Agriculture Organization of the United Nations (FAO, 2008) has enlisted three main areas in which the public can actively play decisive roles in the application of agricultural biotechnology tools: the national policy dialogues that will facilitate the enlightenment of policy makers about the positions, opinions and concerns of different stakeholders should be the first area of active public participation. In Nigeria, activities such as national biotechnology policy formulation by series of public consultations; the formulation of policy statement for novel foods of unfamiliar plant and animals' origin; and resident panels for policy negotiations should be routinely conducted to ensure a wider scope of understanding of agricultural biotechnology by the public. Another area for the involvement of the public is the development of regulatory framework for products of agricultural biotechnology. The consensus highlighted by the 23rd article of the Cartagena protocol (Mackenzie et al., 2003) has enlisted public-engagements, education and participation as the prerequisites for the establishment of biosafety framework which Nigeria already has.

8.4 COMMUNICATING AGRICULTURAL BIOTECHNOLOGY, THE NIGERIA STORY

The use of biotechnology tools to improve agricultural production has been a subject of heated debate especially in Nigeria where the technology has been received with much suspicion. The debate has generated more heat than light, largely because of the limited understanding by the media, policy makers, civil society groups, Religious groups, farmers and other players in the sub-sector who rely on the media for agricultural information. Misinformation remains one of the key factors that have hindered the adoption of agricultural biotechnology to improve farming in Nigeria and Africa as a whole. Yet, the media – namely radio, television and newspapers – have the potential to address this challenge by providing timely and accurate information due to their wide reach and use of local dialects, especially in print journalism, radio, television and workshop/seminars. This chapter sets out to establish how the media, workshops and seminars can be used as an effective tool to relay the correct information to various players in the sector, with Nigeria as a Case Study.

8.5 MEDIA CONTENT ANALYSIS OF THE OFAB NIGERIA CHAPTER IN THE MEDIA FROM 2012 TO 2019

The media content analysis was initiated to gauge the impact of the numerous advocacies and investments of the OFAB Nigeria Chapter and the NABDA in the media from 2012 to 2019. The analysis was to, among other things, determine: (a) if the media had been able to successfully understand the concept of modern biotechnology and its tools including GMOs, the benefit of this technology to the nation as well as the health and wellbeing of Nigerians; if the media is positively contributing to or acting as a stumbling block to the perception of Nigerians on the need for the technology; (b) if there still exist knowledge gaps in the media that can be exploited; (c) if there are inadequacies in the media that OFAB and NABDA need to urgently address and which media organizations in the country are against modern biotechnology and highlight what must be done to reverse the tide.

After months of search at the National Library, a total of 1013 articles were identified and analysed from 13 newspapers, television (YouTube) and radio (Sound Cloud) for 2012 to 2017. There was an unprecedented surge in the number of articles tracked for 2018 to 2019 as a total of 346 articles were recorded from newspapers, while the electronic media tracked via television and radio has over 315 stories. On the average, 253.42 stories were published yearly, 21.119 articles were published on a monthly basis and 3.01 stories on a weekly basis. Among all the articles identified and analysed in the years under review, newspaper had the highest coverage while the radio and television had the least coverage for Biotechnology/GM stories (Figure 8.1). This means that OFAB/NABDA needs to scale up efforts, such as sensitization and capacity building workshops, aimed at getting the media outlets to understand the technology and how it can benefit the country. In addition, deliberate efforts, such as visits to media houses and training of news editors in biotechnology, should be made. More so, efforts should be made to support journalist networks from northern Nigeria; most of the articles analysed were authored by southern-based reporters.

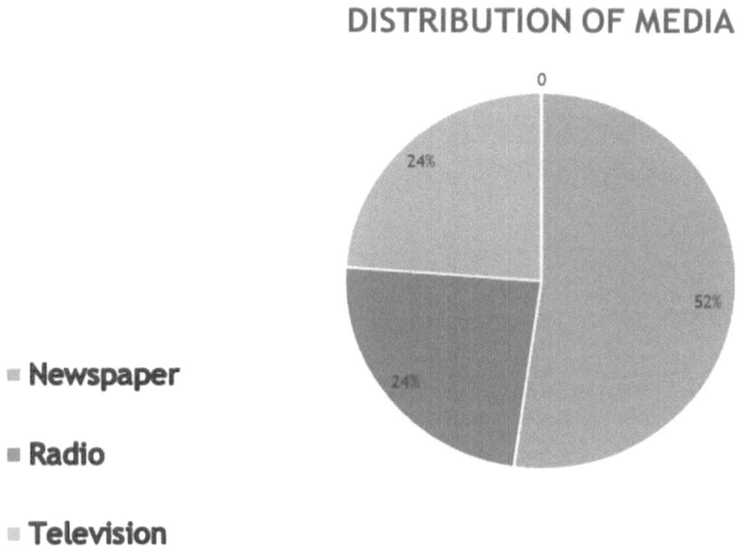

FIGURE 8.1 Biotechnology/GM stories in media outlets.

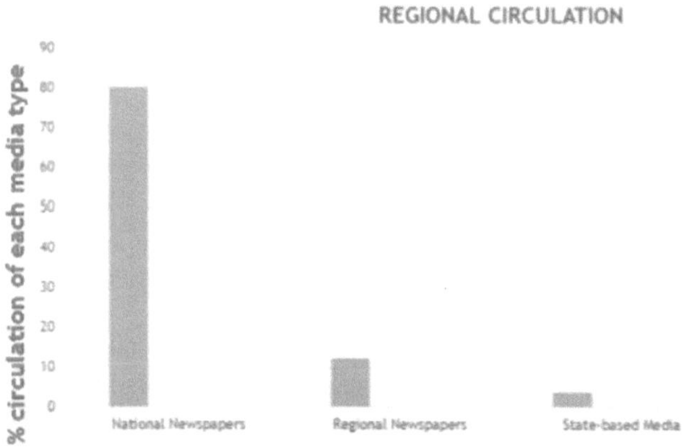

FIGURE 8.2 Circulation of Biotechnology/GM stories in newspapers across Nigeria.

Circulation of Biotechnology/GM stories (Figure 8.2) was, however, higher in national newspapers followed by regional and state-based media being the least. This implies the need for OFAB/NABDA to reach out more to the grassroots by engaging the local press in its events at the state and local levels.

Furthermore, the story focus of all the articles was more of promotion of Biotechnology/GM and less of regulation. This indicates that OFAB/NABDA needs to establish a balance (Figure 8.3).

Additionally, favourable space has the most allocation in Biotechnology/GM stories followed by unfavourable and neutral being the least (Figure 8.4). Moreover, 1–5 inches has the highest allocation in Biotechnology/GM stories followed by 6–10

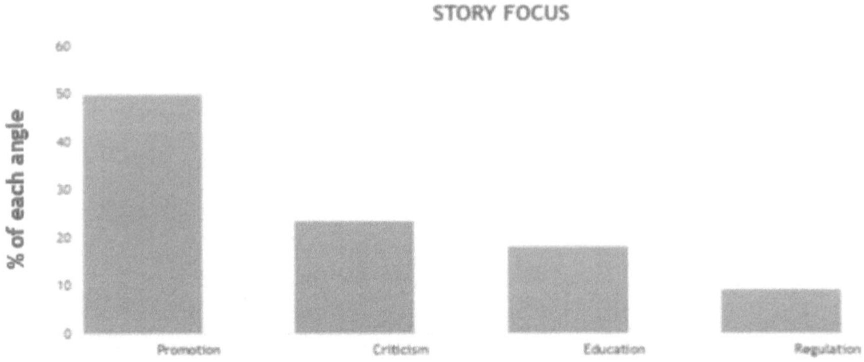

FIGURE 8.3 Biotechnology/GM story angles in newspapers.

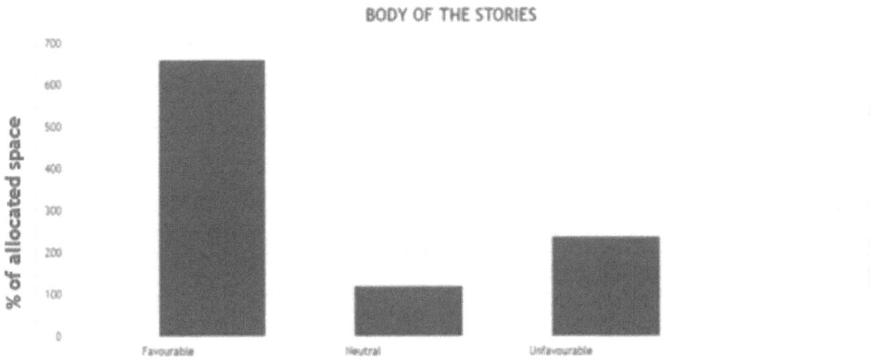

FIGURE 8.4 Space allocated to Biotechnology/GM stories in the pages of newspapers.

inches and above 25 inches being the least (Figure 8.5). This means that OFAB/NABDA needs to work towards getting above 25 inches of space allocation.

Generally, OFAB/NABDA has done a great job in its advocacy and enlightenment programmes especially as it relates to educating the media, taking into consideration the amount of investment and time allotted to issues of the media, but there still exists a knowledge gap in the way issues of modern biotechnology are reported by the media. Conversely, majority of media houses and reporters still see the technology as alien, an imported source of death, one that has no place in the country, a technology that should be avoided, has many unanswered questions and uncertainty surrounding its usage, and above all a technology not accepted by Europe so why should Nigeria accept it. This opinion is strongly propagated by *The Guardian* newspaper which has a number of editorials condemning GMOs. Other media organizations do not express strong views or opinions about GMOs; they have, however, opened their doors to columnists and opinion writers who lack basic knowledge on the subject matter and therefore confuse readers by instilling fear and presenting unsubstantiated claims of what the technology can cause. In this regard, a national workshop should be hosted

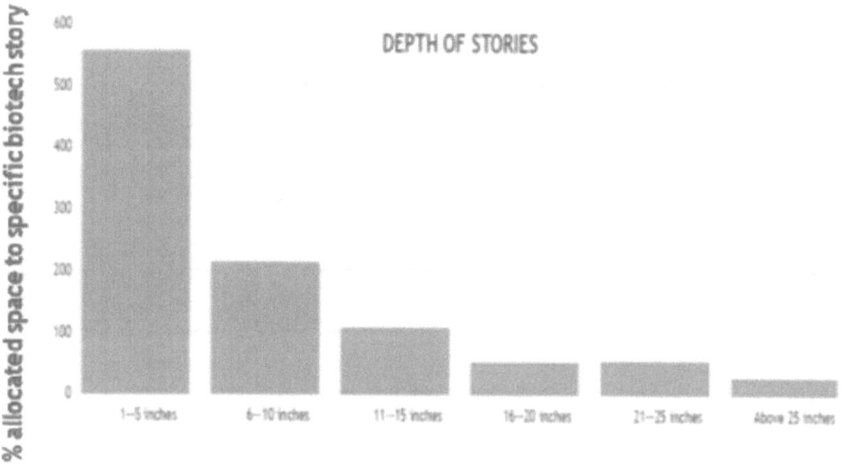

FIGURE 8.5 Space allocated to a particular Biotechnology/GM story on the page of a newspaper.

for editors and columnists that will serve as a platform for OFAB/NABDA to educate them on the technology and help them understand that Nigeria is practising a home-grown version of biotechnology.

8.6 COMPARATIVE ANALYSIS OF THE EVENTS OF OFAB NIGERIA CHAPTER

8.6.1 Primary Data Comparative Analysis for the Years 2011–2017

Generally, the survey conducted in 2011–2013 and 2014–2017 on consumption of GMOs showed that GMOs gained massive acceptance in the society. However, it was also clearly reported that OFAB and Internet were essentially the predominant means of disseminating GMOs information among people, while radio, newspaper and television (TV) appeared to contribute less in communicating GMOs reports to the Nigerian populace. Further survey carried out in 2018 and 2019 also showed an increased consumption and acceptance of GMOs. Described below are three distinct comparative analyses of surveys in relation to professional bodies OFAB events for 2018, OFAB at other outreach programmes for 2018 and OFAB events for 2019.

8.6.2 Comparative Analysis of Professional Bodies OFAB Events for the Year 2018

The comparative analysis involving gender distribution (Figure 8.6), acceptability of GMOs consumption (Figure 8.7), food security (Figure 8.8), use of less chemical sprays (Figure 8.9), sources of biotechnology information (Figure 8.10) and labelling of GM products (Figure 8.11) was reported for OFAB 2018 events. These events

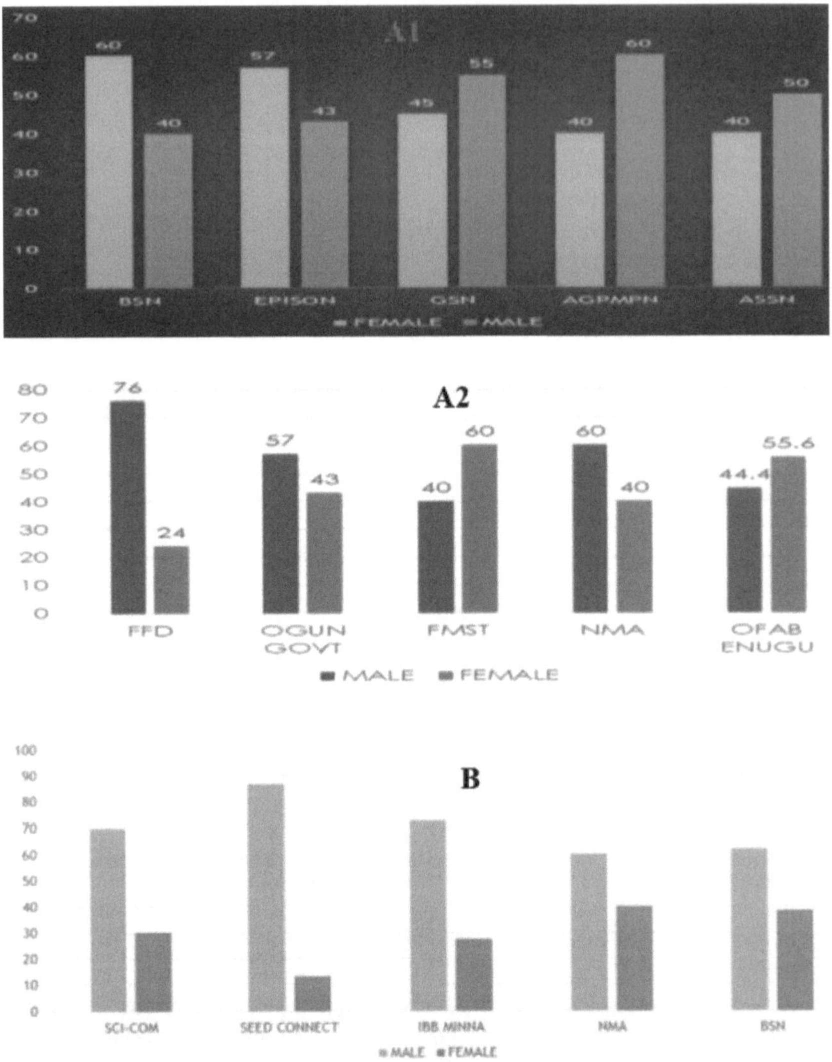

FIGURE 8.6 Distribution of males and females for OFAB events in 2018 (A1 and A2) and 2019 (B). BSN, Biotechnology Society of Nigeria; EPiSON, Epidemiology Society of Nigeria; GSN: Genetic Society of Nigeria; AGPMPN, Association of General and Private Medical Practitioners of Nigeria; ASSN, Association of Seed Scientists of Nigeria; FFD, Farmers Field Day; FMST, Federal Ministry of Science and Technology; NMA, Nigeria Medical Association; SCI-COM, Science Development Network; IBB, Ibrahim Badamasi Babangida University.

involved four professional bodies, i.e., Biotechnology Society of Nigeria (BSN), Genetics Society of Nigeria (GSN), Epidemiology Society of Nigeria (EPiSON), Association of General and Private Medical Practitioners of Nigeria (AGPMPN) and Association of Seed Scientists of Nigeria (ASSN). Accordingly, the result of surveys on distribution of gender of professional bodies in attendance during the events

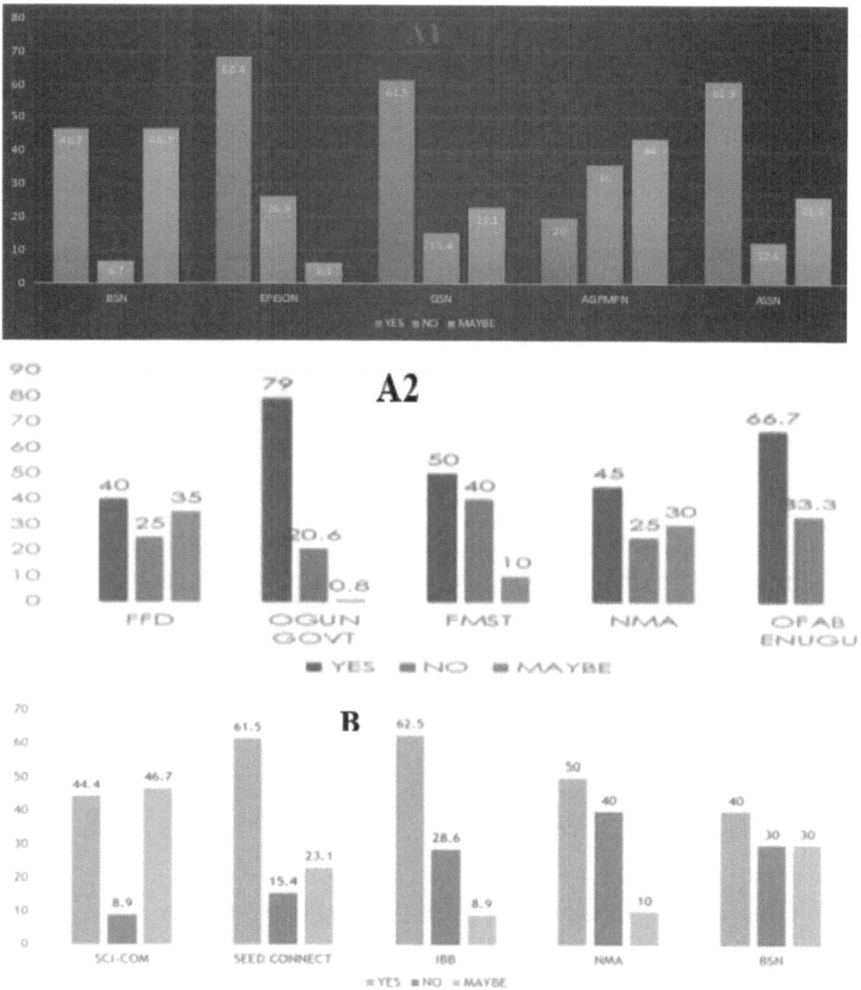

FIGURE 8.7 Opinions of participants on the consumption of GM products during OFAB events in 2018 (A1 and A2) and 2019 (B). BSN, Biotechnology Society of Nigeria; EPiSON, Epidemiology Society of Nigeria; GSN, Genetic Society of Nigeria; AGPMPN, Association of General and Private Medical Practitioners of Nigeria; ASSN, Association of Seed Scientists of Nigeria; FFD, Farmers Field Day; FMST, Federal Ministry of Science and Technology; NMA, Nigeria Medical Association; SCI-COM, Science Development Network; IBB, Ibrahim Badamasi Babangida University.

indicated a close spread of male and female attendees, where AGPMPN, GSN and ASSN had highest attendance of males while BSN and EPiSON had females as the highest attendees (Figure 8.6).

Similarly, the opinion of people on consumption of GMOs was sought from surveys reported in the events and whose results indicated that people will readily consume GM products in Nigeria across board except for the medical group AGMPN

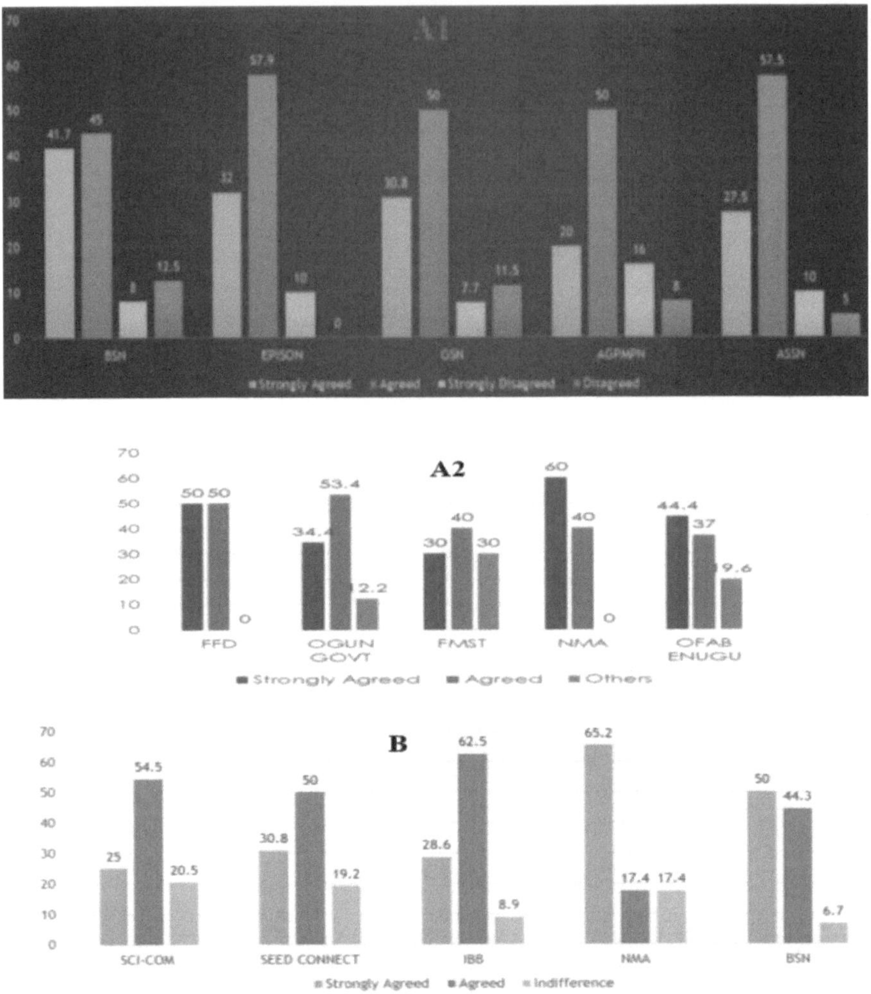

FIGURE 8.8 Opinions of participants on the use of modern biotechnology in addressing food security during OFAB events in 2018 (A1 and A2) and 2019 (B). BSN, Biotechnology Society of Nigeria; EPiSON, Epidemiology Society of Nigeria; GSN, Genetic Society of Nigeria; AGPMPN, Association of General and Private Medical Practitioners of Nigeria; ASSN, Association of Seed Scientists of Nigeria; FFD, Farmers Field Day; FMST, Federal Ministry of Science and Technology; NMA, Nigeria Medical Association; SCI-COM, Science Development Network; IBB, Ibrahim Badamasi Babangida University.

(Figure 8.7). Indeed, it remains a fact that every government plan to attain food security in her country and how this can be addressed is reliant upon the application of biotechnology in agriculture (Figure 8.7).

In a bid to address this, the above professional bodies, including the medical group (AGMPN) and in whose events surveys were carried out, agreed to the use of GM technology in addressing food security in Nigeria (Figure 8.8).

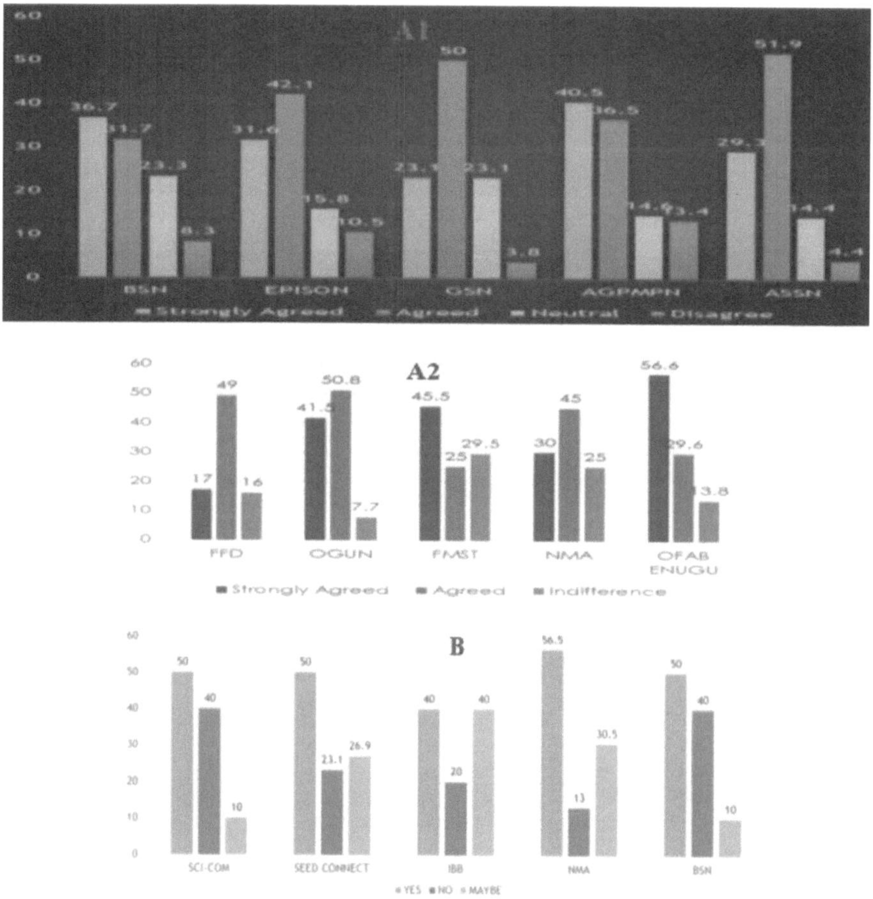

FIGURE 8.9 Opinions of participants on the use of modern biotechnology in reducing the use of chemical sprays by farmers during OFAB events in 2018 (A1 and A2) and 2019 (B). BSN, Biotechnology Society of Nigeria; EPiSON, Epidemiology Society of Nigeria; GSN, Genetic Society of Nigeria; AGPMPN, Association of General and Private Medical Practitioners of Nigeria; ASSN, Association of Seed Scientists of Nigeria; FFD, Farmers Field Day; FMST, Federal Ministry of Science and Technology; NMA, Nigeria Medical Association; SCI-COM, Science Development Network; IBB, Ibrahim Badamasi Babangida University.

Moreover, the use of chemical sprays is now known to be a major hitch in agriculture due to the cost of purchasing the chemicals and the negative impact the chemicals have on famers (Figure 8.9). Surveys on the use of fewer chemical sprays were reported by the four professional bodies, and based on their findings, the use of biotechnology in agriculture was adjudged to be strongly preferred over chemical sprays, and this could be due to the fact that biotechnology has the tendency to reduce much dependence on chemical sprays by farmers (Figure 8.9).

Furthermore, information dissemination and the medium used are crucial in every society and a new technology needs a good medium to disseminate such information

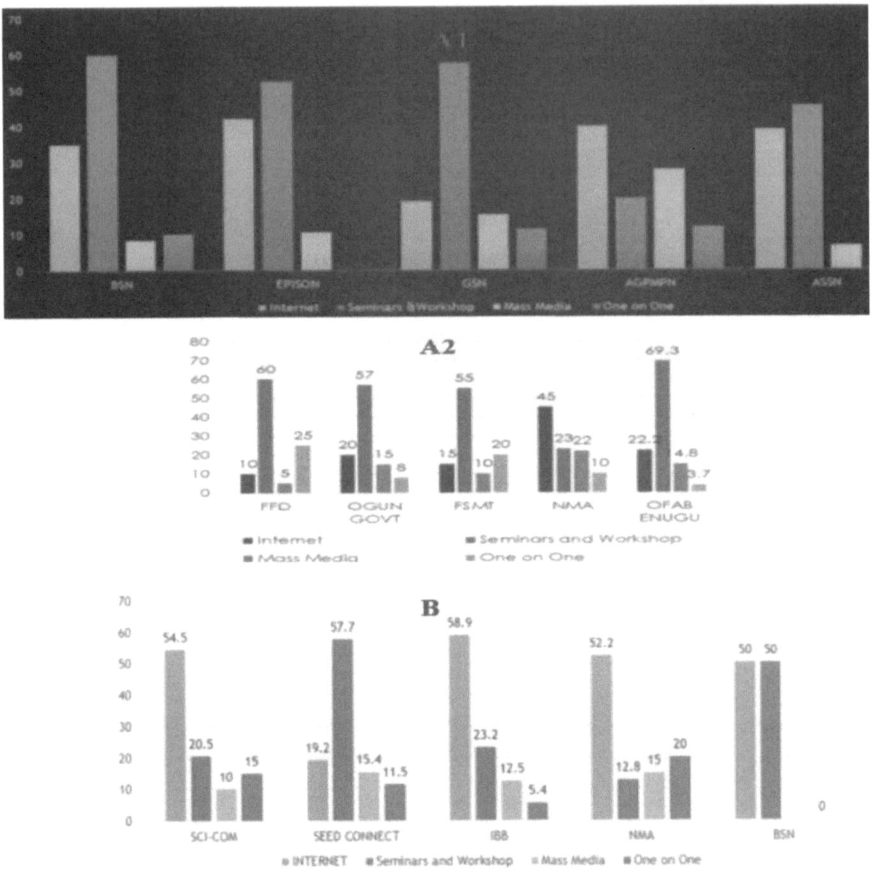

FIGURE 8.10 Opinions of participants on the sources of modern biotechnology information during OFAB events in 2018 (A1 and A2) and 2019 (B). BSN, Biotechnology Society of Nigeria; EPISON, Epidemiology Society of Nigeria; GSN, Genetic Society of Nigeria; AGPMPN, Association of General and Private Medical Practitioners of Nigeria; ASSN, Association of Seed Scientists of Nigeria; FFD, Farmers Field Day; FMST, Federal Ministry of Science and Technology; NMA, Nigeria Medical Association; SCI-COM, Science Development Network; IBB, Ibrahim Badamasi Babangida.

to the would-be users and consumers of the technology (Figure 8.10). In keeping with this fact, a survey aimed to determine the best sources of information on modern biotechnology was reported in the 2018 events, and analysis of findings alludes to internet, seminars and workshops as the three most important sources of communication of the new technology (Figure 8.10).

More so, the labelling of products is very vital in products' purchases, and in order to gain insight into the relevance of labelling GM products, surveys were reported by the four professional bodies, and in the surveys, a lot of the participants agreed to the labelling of GM products not really because they are against it but rather to avail them the opportunity of knowing what the product is made of (Figure 8.11).

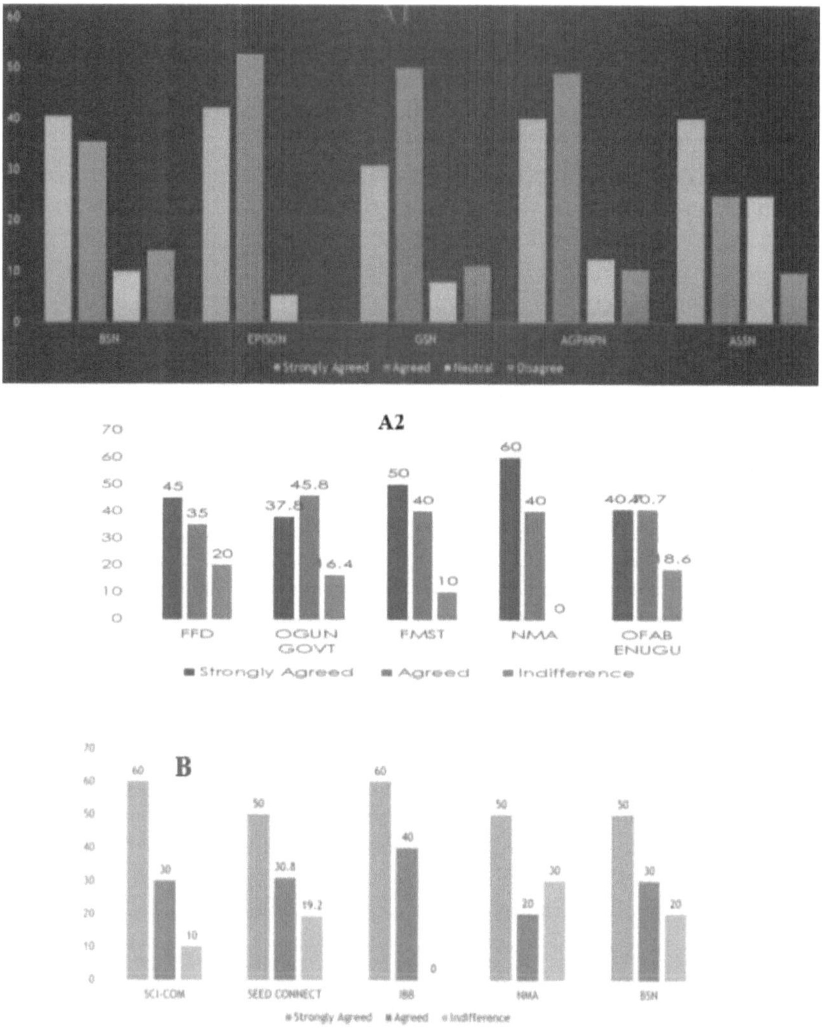

FIGURE 8.11 Opinions of participants on the labelling of GM products during OFAB events in 2018 (A1 and A2) and 2019 (B). BSN, Biotechnology Society of Nigeria; EPISON, Epidemiology Society of Nigeria; GSN, Genetic Society of Nigeria; AGPMPN, Association of General and Private Medical Practitioners of Nigeria; ASSN, Association of Seed Scientists of Nigeria; FFD, Farmers Field Day; FMST, Federal Ministry of Science and Technology; NMA, Nigeria Medical Association; SCI-COM, Science Development Network; IBB, Ibrahim Badamasi Babangida University.

8.6.3 Comparative Analysis of OFAB at Other Outreach Programmes for the Year 2018

The comparative analysis involving gender distribution, food security, sources of biotechnology information, acceptability of GMOs consumption, labelling of GM products and use of fewer chemical sprays were reported for outreach programmes of OFAB in 2018. The affected outreach programmes included Farmer's Field Day (FFD), Federal Ministry of Science and Technology (FMST), Nigerian Medical Association (NMA), Zonal sensitization workshop in Ogun state (Ogun-Govt) and Biosafety and Biotechnology Workshop (OFAB-Enugu). In a bid to address data distribution of gender across the events, surveys on attendance of participants were conducted, and the ensuing reports suggest a close distribution of males and females for the OFAB events in 2018, with exception of FFD which had males as the highest attendees than females. As a way to answering the question "Do GMOs have the ability to address food security in Nigeria?" a report involving the above professional bodies and in whose events surveys were carried out strongly agreed that GM technology could be used in addressing food security in Nigeria. In the same vein, the dissemination of information has proven to be a key factor for the acceptability of every new technology, and on this basis, the major sources of biotechnology information in Nigeria were surveyed and the findings of the reports as per the above outreach programmes suggest internet, seminars and workshops as the three most important sources of the modern technology. In order to understand the degree to which GMOs are accepted for consumption in Nigeria, a survey was reported by the five programmes whose findings, especially those of Ogun-Govt and OFAB-Enugu, suggest that Nigeria's populace will readily consume GM products that have been endorsed by the competent authority. Moreover, the relevance of labelling of products for purchases cannot be over-emphasized. As a way to evaluating the place of labelling GM products in Nigeria, the five outreach programmes reported the outcome of their surveys which show that a lot of the participants agreed to the labelling of GM products not really because they are against it but rather to avail them the opportunity of knowing what the product is made of. Furthermore, the use of chemical sprays is often regarded as a set-back in agricultural practices on account of the cost of purchasing the chemicals and also the negative impact of the chemicals on farmers. In agreement with this fact, the above five programmes reported their findings which show that farmers in Nigeria accept the technology than the use of chemical sprays.

8.6.4 Comparative Analysis of OFAB Events for the Year 2019

The comparative analysis involving gender distribution, food security, sources of biotechnology information, acceptability of GMOs consumption, use of less chemical sprays and labelling of GM products was reported for 2019 OFAB events. These events included Science Development Network (SCI-COM), National Seed Council Seed Connect Conference 2019 (Seed Connect), OFAB Nigeria at the Annual Conference of the Faculty of Science at Ibrahim Badamasi Babangida University (IBB), NMA and BSN. In order to determine gender, spread at the five events, surveys were conducted whose results showed a close distribution of males and females for NMA and BSN events but a wide margin existing for other events like Seed connect, SCI-COM and IBB in which males constituted the major attendees in the events. Similarly, the acceptability of GM products consumption was assessed through surveys during the events, and findings showed that the Nigerian populace will readily consume GM products. The probing of major sources of biotechnology information was also done through surveys involving the five events, and findings suggest internet, seminars and workshops as the three most important sources of biotechnology information. In a bid to ascertain the ability of GMOs to address food security in Nigeria, surveys were conducted during the five events whose findings showed strong agreement of people to the use of the GM technology in addressing food security in Nigeria. In addition, the survey involving people perception on reduction of chemical spray by farmers in farms were also carried out in the events, and the ensuing results showed acceptance of the new technology due to the fact that it will reduce the use of chemical sprays by farmers. The relevance of labelling GM products was also probed through surveys during the five events, and the resulting findings indicate that a lot of the participants agreed to the labelling of GM products not really because they are against it but rather to avail them the opportunity of knowing what the product is made of.

8.7 CONCLUSION

Analysing the various roles of effective communication in the adoption and advancement of agricultural biotechnology in Nigeria cannot be over-emphasized. The SDG of zero hunger by 2030 is fundamental for national development and wellbeing of Nigeria, and the facilitation of adequate understanding of the several ways Nigeria can achieve this goal is very fundamental. Communication of agricultural biotechnology, among other tools, is necessary to enabling Nigeria attain food security. The application of deliberate efforts that will encourage stakeholders to participate in science-based decision is crucial and necessary for the acceptance and adoption of innovative technology such as agricultural biotechnology that can fast track national development in Nigeria.

REFERENCES

AATF. (2017). *The Open Forum on Agricultural Biotechnology (OFAB): A Decade of Success, 2006–2016*. Nairobi, Kenya: African Agricultural Technology Foundation. ISBN 9966-775-13-7.

Aina, P.A., Patil, A., Ejembi, P., Oloketuyi, S. and Aina, I. (2019). Scientific Advancement in Contemporary Society: The Interface between Technology, Science and Law. *European Scientific Journal* 15; DOI: 10.19044/esj.2019.v15n32p127.

Arntzen, C., Coghlan, A., Johnson, B., Peacock J. and Rodemeyer, M. (2003). GM Crops: Science, Politics and Communication. *Nature Reviews* 4: 838–843.

Arthur, G.D. (2011). Benefits and Concerns Surrounding the Cultivation of Genetically Modified Crops in Africa: The Debate. *African Journal of Biotechnology* 10(77): 17663–17677.

Burns, T.W., O'Connor, D.J. and Stocklmayer, S.M. (2003). Science Communication: A Contemporary Definition. SAGE Journals Online. *Science* 12: 183–202.

Castillo, G. (2003). *Science Communication Whose Time Has Come: An Evaluation Report on the Global Knowledge Center on Crop Biotechnology*, Los Baños, Laguna, Philippines: International Service for the Acquisition of Agri-Biotech Applications (ISAAA). 89pp.

Devos, Y., Maeseele D., Rehuel L., Van S. and Dewaele, D. (2007). Ethics in the Societal Debate on Genetically Modified Organisms: A (Re)quest for Sense and Sensibility. *Journal of Agricultural and Agricultural Ethics* 31(1): 29–61

Einsiedel, E. (2008). Introduction: Making Sense of Emerging Technologies. In *First Impressions: Understanding Public Views on Emerging Technologies*. E. Einsiedel (Ed.). Genome Prairie: University of Calgary, Canada, pp. 2–4.

Ezezika, O., Daar, A., Barber, K., Mabeya, J., Thomas, F., Deadman, J., Wang, D. and Singer, P. (2012). Factors Influencing Agbiotech Adoption and Development in Sub-Saharan Africa. *Nature Biotechnology* 30: 38–40.

Food and Agriculture Organization of the United Nations (FAO) (2008). Public Participation in Decision-making Regarding GMOs in Developing Countries: How to Effectively Involve Rural People. Electronic Forum on Biotechnology in Food and Agriculture: Conference 11. Retrieved from the World Wide Web: http://fao.org/biotech/C12doc.htm.

Gil, J.D.B., Reidsma, P. and Giller, K. (2019). Sustainable Development Goal 2: Improved Targets and Indicators for Agriculture and Food Security. *Ambio* 48: 685–698. DOI: 10.1007/s13280-018-1101-4.

Hallman, W. K., (2008). GM foods in hindsight. In E. Einsiedel (Ed.). *Emerging Technologies: From Hindsight to Foresight*. University of British Columbia Press, pp. 13-32.

Holliman, R., Whitelegg, E., Scanlon, E., Smidt, S. and Thomas, J. (2009). *Investigating Science Communication in the Information Age: Implications for Public Engagement and Popular Media*. Oxford University Press.

Irwin, A. (2009). Moving Forwards or in Circles? Science Communication and Scientific Governance in an Age of Innovation. In *Investigating Science Communication in the Information Age: Implications for Public Engagement and Popular Media*. R. Holliman, E. Whitelegg, E. Scanlon, S. Smidt, and J. Thomas (Eds.). Oxford University Press, pp. 1–17.

Jennifer, K. (2016). Policy: Reboot the Debate on Genetic Engineering. *Nature. International Weekly Journal of Science* 531 (7593): 165–167, DOI: 10.1038/531165a.

Jonathan, V.S. (2017). Science Communication: A Science Not an Art. *Analytical Chemistry* 89: 2645–2645.

Leshner, A.I. (2003). Public Engagement with Science. *Science* 299(5609): 977.

Letrud, K. and Hernes, S. (2019). Affirmative Citation Bias in Scientific Myth Debunking: A Three-in-One Case Study. *PLoS One* 14(9): e0222213. DOI: 10.1371/journal.pone.0222213.

Mackenzie, R., Burhenne, G.F., La Viña, A.G.M., Werksman, J.D., Ascencio, A., Kinderlerer, J., Kummer, K. and Tapper, R. (2003). *An Explanatory Guide to the Cartagena Protocol on Biosafety*. Gland, Switzerland and Cambridge, UK: IUCN. XVI +295pp.

Mariechel, J.N. (2003). *Science Communication: Building Consensus on Crop Biotechnology*. Adapted from: https://webcache.googleusercontent.com/search?q=cache:2W81jsWh3dUJ:https://www.isaaa.org/resources/publications/communication_challenges_and_convergence_in_crop_biotechnology/download/communication_challenges_and_convergence_in_crop_biotechnology-chapter2.pdf+&cd=1&hl=en&ct=clnk&gl=ng&client=opera.

Michael, C., Falk, B.M., Chassy, S.K., Harlander, T.J., Hoban, I.V., Martina, N.M. and Amin, R.A. (2002). Food Biotechnology: Benefits and Concerns. *The Journal of Nutrition* 132(6): 1384–1390.

National Academies of Sciences, Engineering, and Medicine (2017). Using Science to Improve Science Communication. In *Communicating Science Effectively: A Research Agenda*. Washington, DC: The National Academies Press. DOI: 10.17226/23674.

Navaro, M.J. (2011). *Communication Challenges and Convergence in Crop Biotechnology*. Ithaca, New York: International Service for the Acquisition of Agri-Biotech Applications (ISAAA) and Philippines: SEAMEO Southeast Asian and Regional Center for Graduate Study and Research in Agriculture (SEARCA). Chapter 2, pp. 26–42.

Navarro, M., Natividad, K., Mercado, S. and Panopio, J. (2013). Visual Representation of Science: How Cartoonists Define Crop Biotechnology. *International Journal of Current Research* 5(2): 668–674.

Peters, H.P. and Sawicka, M. (2007). *German Reactions to Genetic Engineering in Food Production. In the Public, the Media and Agricultural Biotechnology.* D. Brossard, J. Shanahan and T.C. Nesbitt (Eds.). Oxon: CAB International, pp.57–96.

Poortinga, W. and N. Pidgeon. (2007). *Public Perceptions of Agricultural Biotechnology in the UK: The Case of Genetically Modified Food. In the Media, the Public and Biotechnology.* D. Brossard, J. Shanahan and T.C. Nesbitt (Eds.). Oxon: CAB International, pp. 21–56.

Rowe, G. and Frewer, L. (2004). Evaluating Public Participation Exercises: A Research Agenda. *Science, Technology and Human Values* 29: 4–29. 10.1177/0162243903259197.

Sigrid, B. (2013). *Is Green Genetic Engineering Sustainable?* alumniportal-deutschland.org Adapted from: https://www.alumniportal-deutschland.org/en/global-goals/sdg-02-hunger/green-genetic-engineering/.

Stilgoe, J. and Wilsdon, J. (2009). The New Politics of Public Engagement with Science. In *Investigating Science Communication in the Information Age: Implications for Public Engagement and Popular Media.* R. Holliman, E. Whitelegg, E. Scanlon, S. Smidt and J. Thomas (Eds.). Oxford University Press, pp. 1–17.

The International Service for the Acquisition of Agri-biotech Applications. (2020). *Communicating Crop Biotechnology. Pocket K No. 33.* Ithaca, NY: ISAAA. Adapted from: https://www.isaaa.org/resources/publications/pocketk/document/Doc-Pocket%20K33.pdf.

Uche, C. and Familusi, L.C. (2018). Topical Review of Hidden Hunger and Poverty Profile in Nigeria: Case Study of Kwara State (Province) and Suggested Agricultural and Economic Solutions. *Global Journal of Management and Business* 18(4) Version 1.0.

United Nations. (2015). *Population 2030: Demographic Challenges and Opportunities for Sustainable Development Planning (ST/ESA/SER.A/389).* Adapted from: https://www.un.org/en/development/desa/population/publications/pdf/trends/Population2030.pdf.

United Nations. (2017). *Goal 2: Zero Hunger.* Accessed on the 9th April, 2020. Adapted from: https://www.undp.org/content/undp/en/home/sustainable-development-goals/goal-2-zero-hunger.html.

Walker, S. (2011). *Biotechnology Demystified: A Self-Teaching Guide.* McGraw Hill, 276p.

9 Current Status of Biotechnology Manpower Development in Nigeria

James C. Ogbonna
University of Nigeria

Benjamin Ewa Ubi
Ebonyi State University
Tottori University

Abdulrazak Ibrahim
Ahmadu Bello University

Ebiamadon Andi Brisibe
University of Calabar

Abubakar Gidado, Mwajim Bukar, and Aliyu Daja
University of Maiduguri

CONTENTS

DOI: 10.1201/9781003179177-9

9.1 INTRODUCTION

As we stand at the beginning of a transition to a bioeconomy, disruptive technologies including several biological research innovations will be the most fundamental tools required by any society – especially a developing country like Nigeria – for the creation of wealth and jobs to develop a sustainable economy for a rapidly increasing population. The advent of breakthrough technologies such as biotechnology and its global impact has shown that "brain-power" and not matter is fast becoming the most valuable of human assets. Interestingly, some of these technologies, especially the development of new materials including biodegradable plastics and new energy systems such as bio-based production of chemicals and fuels including fuel ethanol and biodiesel, are scale neutral and can thus help people both in developing and developed societies to leapfrog industrial development. To address the specific needs of our national challenges in food security, health, energy, poverty alleviation and proper environmental and natural resource management, for example, Nigeria must become an active participant in the development of biotechnology.

What is most exciting about the science of biotechnology in a developing economy such as Nigeria is the range of possibilities for what is yet to come from this vibrant area of research. The field is extremely broad and multi-disciplinary, with applications ranging from energy and biomaterials production to product development in pharmaceuticals and molecular and tissue engineering, solving problems of land and water pollution to developing high-yielding crop varieties and tools that can provide the impetus for increased and sustainable food production per annum that would help to raise the average dietary energy intake of the people. Biotechnology has a centrally unique role with great untapped potential to address pressing local and global challenges including climate change and sustainable growth and development. It has a great deal to offer most developing countries, long promising the prospect of adapting agricultural crops and livestock to entirely new purposes. For instance, the 21st-century medicine promises to utilize the potential of biotechnology to conquezr preventable diseases and cancer, grow new blood vessels in the heart, create new organs from stem cells, develop personalized medicines and perhaps even prevent the entire process of ageing. Another cutting-edge research area is the integration of big data and artificial intelligence with biotechnology, which holds a lot of promises.

In spite of the exciting possibilities of rapid economic growth and research opportunities offered in the different aspects of advanced biotechnology research, Nigeria still seems to be inured in the midst of an "arrested development". Invariably, the biotechnology sector is knowledge-driven and requires highly skilled manpower. For Nigeria to take advantage of these biotechnology potentials, there is a need to adopt robust biotechnology education programmes starting from secondary school to create awareness and instil interest in the students to choose career in biotechnology, to undergraduate level to provide basic biotechnology education, to postgraduate education to produce experts in various areas of biotechnology. The formal biotechnology education at the different levels of educational system should be supplemented with informal hands-on practical training for mastery of the various techniques used in biotechnology research and development.

9.2 EVOLUTION OF BIOTECHNOLOGY IN NIGERIA

The traditional forms of biotechnology started as old as the human civilization and its applications such as in cheese making, bakery, wine making, etc. were in common place dawn of the ages. In the late 1970s, bio/technology (biological technology) emerged as a new discipline with perspectives rapidly moving into its modern forms of application in continuum with the traditional forms. Nigerian scientists recognized the great importance of this emerging scientific discipline leading to the formation of the Biotechnology Society of Nigeria (BSN) at the beginning of the 1980s as a professional body for all biological scientists and the related subject matter specialists to come together and exchange ideas towards promoting biotechnological advancements in Nigeria and in cooperation with the international community. Early in its formation, the society held the firm belief that the proper development and utilization of biotechnological processes in the provision of goods and services would provide the panacea for rapid economic development in Nigeria.

In 1983, United Nations Industrial Development Organization (UNIDO) established the International Centre for Genetic Engineering and Biotechnology (ICGEB) in Trieste and New Delhi. Nigeria was accorded the status of an affiliate member of ICGEB. To further boost the country's yearnings for biotechnological advancements and utilization, the National Centre for Genetic Resources and Biotechnology (NACGRAB) was established in Ibadan, Nigeria in 1987. In the early 1990s, the then government provided funding for the development of the first Master Plan for Biotechnology in Nigeria; with the provided seed money enabling the National Agency for Science and Engineering Infrastructure (NASENI) carry out "Needs Assessment" of designated universities in Nigeria in collaboration with the Israeli embassy in Lagos aimed at ascertaining the status of biotechnology in Nigeria. The Needs Assessment was followed by the first ever (government-supported) National Conference on Biotechnology held in Lagos (Dubar Hotel) in 1993, which was remarkably chaired and declared open by His Excellency, General I. Babangida (Rtd.). Thereafter, NASENI embarked on capacity building, with support from the United Nations Scientific and Cultural Organization (UNESCO), to raise critical mass of scientists to drive this cutting-edge technology in Nigeria. In the same year (1993), Nigeria established the Sheda Science and Technology Complex (SHESTCO) which also houses the Biotechnology Advanced Laboratory.

In the year 2000, a Committee was inaugurated by the then Honourable Minister of Science and Technology to draft the Biotechnology Policy for Nigeria. The then government of Chief Olusegun Obasanjo, convinced that biotechnology alongside information and communication technology would drive the 21st century's global and regional economies, endorsed and approved the National Biotechnology Policy in April 2001 that gave rise to the establishment of the National Biotechnology Development Agency (NABDA) (https://nabda.gov.ng/) in November 2001, with the mandate of meeting the national aspirations of Nigeria through the deployment of biotechnology. Through these established government agencies and the pivotal role of the Nigerian experts in the Diaspora and the BSN, biotechnology activities became greatly accelerated with concerted efforts in building the necessary manpower through biotechnology education and professional certification.

9.2.1 Education in Secondary Schools

Presently, biotechnology is not studied as a subject in secondary schools. Rather, related subjects like biology, agriculture, health science, chemistry, physics, mathematics and other STEM subjects are taught. However, there is a need to introduce programmes that will create awareness on the fields of biotechnology as part of STEM education. Also, the Biotechnology Students Society of Nigeria – BSSN (an affiliate of BSN), embracing the need for secondary school students to appreciate the different areas of biotechnology and its importance to the socio-economic development of Nigeria, has been involved in making secondary school students to develop interest in biotechnology as they proceed to tertiary education. It is important that such efforts are encouraged by policy and supported by all universities with existing BSSN. The concept of "catch them young" is very important, and universities, especially those offering undergraduate programmes, should formalize and expand their impact at the secondary education level to facilitate, among others, the recruitment of bright and ready students into their departments and programmes.

9.3 BIOTECHNOLOGY EDUCATION AT UNDERGRADUATE LEVEL

Biotechnology emerged as a new discipline in the late 1970s, albeit its historical practice dawn the ages. The BSN founded in 1981 at the University of Jos was conscious of the need to raise a critical mass of manpower towards the deployment of this burgeoning technology in Nigeria. At the time of its founding in the 1980s, there were no full-fledged Departments of Biotechnology for training undergraduate students in Nigeria. However, in keeping with the society's philosophy of integrating scientists from all areas of biological sciences and all the related subject matter specializations, biotechnology experts were spread in different traditional disciplines of the biological sciences, agriculture, medical and pharmaceutical sciences, environmental sciences, engineering, computer sciences, etc. This spread of experts from various fields in line with the multi-disciplinary nature of biotechnology was adequate at the beginning stages that could be called the "first generation" phase of experts training in biotechnology. However, these traditional life science fields and other associated Faculties did not fully cover all the aspects expected of a more holistic biotechnology education in Nigeria and offered small numbers of discipline-related biotechnology courses in their syllabus which were not enough to adequately train graduates that will lead Biotechnology Research and Development in Nigeria in line with the global trends.

There were initial controversies among scientists on the necessity or otherwise of the B.Sc. or B.Tech. biotechnology programmes in Nigerian Universities. While many scientists argued in favour of undergraduate biotechnology programmes, other scientists favoured biotechnology (and especially molecular biology) being taught as a course(s) at the undergraduate level and then later presented as a specialty at the postgraduate degree level in the associated Faculties of Agriculture, Biological Sciences, Medical Sciences, Pharmaceutical Sciences, Environmental Sciences, etc. In the early 2000s, the need for training undergraduates in full-fledged biotechnology programmes became evident in order to enable more dedication and a holistic approach to biotechnology education, encompassing its different branches, and serve to detach biotechnology from the narrow prism of a one-sided approach inherent in the respective traditional specialties while keeping in view its highly interdisciplinary nature. Thus, the felt-need arose to train scientists with broad-based biotechnology knowledge who can appreciate and apply multi-disciplinary approaches in solving problems related to biotechnology and contribute meaningfully to its application in healthcare delivery, food and nutrition security, sustainable environmental management, industrial revolution, job/wealth creation and economic development. Undergraduate biotechnology education has assumed an increasing global trend and exists in several countries of the globalized world including the United States, Canada, United Kingdom, Japan, India, etc. India started undergraduate biotechnology programmes about the same time as Nigeria (in the early 2000s) and a large number of universities, as well as affiliated private colleges have embraced this biotechnological educational model (Ashok Pandey, Pers. Comm.)

In line with the felt-need for a holistic training that enables the students cross the one-sided limitation of biotechnology specializations, the first three full-fledged B.Sc. biotechnology programmes in Nigeria were established at the University of Calabar (Dept. of Genetics and Biotechnology established in 2001; and gained first accreditation in 2006), Ebonyi State University in 2003 (first accredited in 2008 in the then Dept. of Biochemistry/Biotechnology; and later demerged to a full-fledged Dept. of Biotechnology in 2009), as well as Caritas University, Enugu, in 2004. Other notable universities in the early phase of undergraduate biotechnology education in Nigeria included the Federal University of Technology, Owerri (B.Tech. Biotechnology, 2003); Godfrey Okoye University, Enugu (B.Sc. Biotechnology, 2009); Modibbo Adama University, Yola (B.Tech. Biotechnology, 2010); etc. Since then several other universities have joined in the establishment of B.Sc. (or B. Tech.) biotechnology programmes either as full-fledged departments (which encouraged the programmes to grow rapidly) or embedded in a larger department to be demerged at a later stage.

According to the 2020/21 JAMB Brochure (http://ibass.jamb.gov.ng/assets/uploads/2018-09-03-221207-sciences.pdf), the departmental nomenclatures for the B.Sc. or B.Tech. undergraduate programmes ranged from Biotechnology (over 25 universities), Genetics and Biotechnology (as in University of Calabar and Akwa Ibom State University), Microbiology and Biotechnology (as in Afe Babalolo University, Ado-Ekiti), Microbiology and Industrial Biotechnology (as in Caleb University), Applied Biology and Biotechnology (as in Wesley University) and Biotechnology and Applied Biology (as was formerly at Godfrey Okoye University, Enugu). Fortuitously, Botany Departments in several universities faced with declining student enrolment, decided to change their names into mainly Plant Science and Biotechnology, while keeping their curricula more or less similar to Botany. Several other universities in Nigeria have also recently applied themselves to the creation of Biotechnology Departments including more recently Taraba State University, Jalingo, which admitted her first set of B.Sc. Biotechnology Students in the 2019/2020 session, with a Google search indicating over 120 universities in Nigeria offering B.Sc. or B.Tech. Biotechnology Programme in 2020 (https://fatherprada.com/the-list/119-list-of-nigeria-universities-offering-biotechnology/)

The background specialization of the experts who drafted the curricula for the emerging biotechnology programmes influenced a great deal of the curricula for the available programmes thereby tilting it largely towards the specialty of the respective curriculum developers (e.g. Microbiology, Botany, Plant Science, Biochemistry). Unfortunately, this scenario makes the curricula used by these universities for the B.Sc. or B.Tech. Biotechnology Programme to vary widely, with some of the curricula being highly biased and inclined depending on the department or faculty housing the programme. Besides the former Departments of Botany that largely continued with their curriculum in spite of the fortuitous change in name to Plant Science and Biotechnology, the curricula of the other departments/universities do not have the required consensus for the biotechnology programme. There is therefore the need to quickly harmonize and fill up the observed gaps in the biotechnology curriculum to allow for consensus in the Benchmark Minimum Academic Standard (BMAS) document during programme accreditation exercise across universities in Nigeria. The BSN was conscious of the fact that the National Universities Commission (NUC) developed a "minimum academic standard" or the "Benchmark" for biotechnology programmes in Nigerian Universities; this Benchmark should be seen as minimum and universities may need to go beyond this minimum, as may be necessary.

The BSN realized the need for a review of the available NUC Biotechnology curriculum and had made effort of reaching out to NUC and presented a proposed more holistic four-year undergraduate curriculum for Biotechnology in Nigerian Universities in 2018. As a professional academic body having all the major stakeholders in Biotechnology in Nigeria as members, the society considered it expedient to contribute to the development of a sound undergraduate curriculum for Biotechnology manpower training in Nigerian Universities. BSN mandated a committee made up of representatives from the various universities then running the B.Sc. or B.Tech. Biotechnology and other academics with vast knowledge in biotechnology to develop the curriculum. It is worth noting the fact that a curriculum developed by a professional society made up of all-rounded experts in different aspects of biotechnology would have far-reaching benefits than the output of selected individuals who may have bias towards their specific disciplines. Recently, too, the BSN created the biotechnology education and Curriculum Development Working Group to, among other functions, liaise with the NUC and other relevant stakeholders to expand the frontiers of knowledge and strategy for biotechnology education and Curriculum Development in Nigeria.

The proposed BSN curriculum by the society presented to NUC was therefore meant to fill the observed gaps in the NUC BMAS and ensure coverage of all aspects of biotechnology without bias to any particular field and the production of all-round graduates with the option of specialization in any area of biotechnology at the postgraduate level. The BSN is continuing to work with the NUC for next steps towards a harmonized curriculum for undergraduate biotechnology education in Nigerian Universities. Furthermore, it is the impression of the BSN team that there should be no specialization (options) at the undergraduate level to allow students apply themselves wholly to the courses (apart from some elective courses of their choice and their final year project areas) and later go into their areas of choice for specialization at the postgraduate level.

In most universities, Department of Biotechnology is housed in the Faculty of Biological Sciences (Science/Natural Science/Life Sciences) and they run a four-year degree programme, while some other universities (e.g. Federal University of Technology, Owerri) run a five-year degree programme. If the Biotechnology Programme were to be housed in the Faculty of Agriculture and Forestry, the duration will be expected to be five years as with other programmes in the faculty. Biotechnology should be treated as a professional course, and in order to accommodate most of the relevant courses and have the required six months for industrial training under the Students Industrial Work Experience Scheme (SIWES), a five-year degree programme could be necessary. Nevertheless, a four-year programme as obtainable in other countries can still be adjudged adequate provided most of the relevant courses are covered: this will require the students in the four-year degree programme having to take more courses in each semester than their counterparts in a five-year degree programme.

It must be pointed out that there is basically nothing wrong with the curricula of Nigerian Universities. With the expected review of the proposed undergraduate biotechnology curriculum by the NUC along the lines proposed by the BSN, the implementation of this envisaged harmonized curriculum for undergraduate training in biotechnology should be done in such a way that classroom teaching is complemented with adequate hands-on practical laboratory training. Well-equipped functional laboratories must be provided and manned by competent personnel to expose the students sufficiently to the practical components of their training. The six-month Students Industrial Work Experience Scheme (SIWES) should be strategically planned in such a way that the trainees are affiliated with the relevant industries/laboratories during this period of internship, and academic calendars made favourable for a six-month SIWES.

In order to ensure compliance with adequate training and skills-set required of biotechnology graduates (just like with other professional societies, e.g. Medical Lab. Science Council of Nigeria, Animal Science Council of Nigeria, etc.), there is an urgent need for a Biotechnology Council of Nigeria that will, inter alia, oversee the accreditation of the biotechnology programmes in Nigerian Universities and conduct national certifying exams for qualification as registered biotechnologists after graduation. Similar to other professional disciplines like Accounting, the role of recognized international accreditation bodies will help to ensure and expand the quality and value of Biotechnology graduates.

9.4 BIOTECHNOLOGY AT POSTGRADUATE LEVELS

Postgraduate programmes must aim at attracting very intelligent and bright students with the right aptitude to create and sustain a world-class human capital base in biotechnology in the country. Just a few decades ago, advanced biotechnology research was centred primarily in the USA, Europe and a few Asian countries, and many researchers and students from Nigeria went to these countries to pursue postgraduate studies. However, interestingly the biotechnology landscape in Nigeria has changed tremendously. Within the last 20 years or so, human resource development in biotechnology accelerated tremendously such that local capacity is now available for training and development of researchers and students at the postgraduate level within the country. Presently, although the number of universities offering undergraduate degrees in biotechnology is still limited, almost all the universities in Nigeria, especially the public universities and some private universities, offer postgraduate programmes in various Departments in the Biological, Agricultural, Medical, Pharmaceutical and Health Sciences.

Like most other branches of learning, the postgraduate programme in biotechnology at these universities is aimed at providing students with the skills, knowledge and experience that are needed to pursue a successful career in the field. Presently, postgraduate biotechnology research is undertaken at the levels of Diploma, Masters' and Doctorate degrees and focuses on the adaptation and application of biological processes for commercial and industrial uses. Usually, the duration of the postgraduate programme is one year for the Diploma, between one and a half to two years for the Masters' and approximately three to five years to obtain a Doctorate degree. Postgraduate training is suitable and usually reserved for candidates that have graduated with a Bachelor's degree in science or an allied subject and a strong background in the agricultural and biological sciences including crop science, Botany/Plant Science, animal science, genetics, biochemistry, microbiology, medical science, pharmaceutical science and other related fields, who wish to extend their knowledge and skills for a career in biotechnology research. At the moment, it appears each of the Nigerian Universities pursues independent research directions. However, collective interest appears to be aggregated on the four key areas of animal and plant genetics and breeding, bio-manufacturing of chemicals and biomaterials for various industries including health, food, energy, and agriculture, diagnostics and biomedical technologies, and environmental management and bioremediation.

In the development of their curricula at the postgraduate levels, most of the universities have taken into cognisance the need to ensure that the appropriate courses are taught. Some of the basic courses include biochemistry, microbiology, cell biology, molecular biology, plant/animal and microbial genetics, applied physics and instrumentation, laboratory techniques, infection and immunity, bioinformatics and biostatistics, genetic engineering, plant biotechnology and crop improvement, biodiversity and technologies, nanotechnology, natural products, virology, fermentation and biochemical engineering, protein engineering, laboratory project, genetic technology, immunodiagnostics, intellectual property rights, etc. Aside from these basic courses, there are many courses in the various applied areas of biotechnology, namely, medical biotechnology, industrial biotechnology, food and agricultural biotechnology and environmental biotechnology. The overriding principle behind teaching these courses to postgraduate students, essentially, appears to be the need to endow graduates with an articulate knowledge and sound understanding of the concepts and techniques of biotechnology in order to meet critical needs of society with particular reference to all aspects of human existence, being altered by new biotechnologies.

There is no doubt that growth in any new technology, and especially one as complex as biotechnology, would require a synergistic collaboration between industry and academia, each having to offer its areas of comparative expertise to the union. The industrial sector should define and put forth the needs of the society, while academia would simply need to develop technologies to meet those demands. Unfortunately, there are no robust collaborations between industry and academia with industry seeking novel technologies from academia for translation and in turn advising the academics on the needs of the society to enable the academia offer solutions to meet the needs of industry. This is a rather unfortunate situation. In other climes, for instance, in the USA, Europe and parts of Asia, various research laboratories have been established in collaboration between industry and academia, which is an indication of a rapidly growing alignment between the two. Nigeria should take a cue from this and develop a healthy infrastructure that would support the development of biotechnology both within and outside academic campuses to attract many intelligent young graduates to the field that would help develop biotechnology startups in the country.

9.5 THE ROLE OF WORKSHOPS AND CAPACITY BUILDING IN BIOTECHNOLOGY TRAINING

Capacity building in terms of physical infrastructural development and human resource development is very crucial in biotechnology education. Biotechnology programmes are now offered by many Nigerian Universities at both undergraduate and postgraduate levels. However, the big question remains whether these universities have adequate capacity and resources to produce the much-needed workforce that will pilot Biotechnology activities in Nigeria and beyond.

Lack of physical infrastructure in terms of research equipment, internet access, power and even water supply is a fundamental problem facing the nation. Biotechnology education is expensive since most of the equipment needed for training, research and development are expensive to purchase and maintain in Nigeria. Some of them are delicate and require constant and stable (in terms of voltage output) power supply (at least during operation). Furthermore, since they are imported items and the producers do not have Service Centres in Nigeria, troubleshooting, regular maintenance and repair are very difficult and expensive. Consequently, even the few available ones are often packed once they develop even minor faults. The paucity of these equipment therefore hampers hands-on training. The result is that many universities graduate students without offering them any practical training in basic biological techniques.

The paucity of infrastructure is not limited to tertiary institutions. Even most of the biotechnology-related research institutes do not have most of the basic biotechnology equipment. Furthermore, internship training, whereby undergraduate students spend a period of six months in Biotechnology companies, is a veritable part of undergraduate biotechnology curriculum. Unfortunately, there are very few biotechnology companies in Nigeria to absorb the increasing number of biotechnology students in the country during their industrial training period and after graduation. In addition, there is still an urgent need to equip our tertiary institutions and biotechnology-related research institutes with the basic biotechnology equipment to ensure that our graduates derive the most benefit and training. Internet bandwidth is very low in most of the universities and most of the students do not have access to internet. Constant and adequate water supply to our laboratories as well as constant power supply are both crucial and should, as a matter of urgency, be addressed.

Often a time, lack of these research/training facilities is emphasized without addressing the issue of human resources. The truth is that many of our tertiary institutions lack qualified and experienced academic staff (both teaching and technical staff) to provide quality biotechnology education. Some of the universities may have adequate numbers, but some of the staff do not have appropriate training and experience, while there are no staff in some of the core areas. Furthermore, there is general shortage of technical staff (laboratory technologists and technical officers) to operate and maintain the few biotechnology equipment in their institutions. Evidence abound that some institutions purchased some equipment that were never installed and used because of lack of qualified personnel to man them. There is therefore a need to recruit and train academic and technical staff for efficient service delivery.

Presently, the gap created by inadequate training facilities and qualified staff in some of the tertiary institutions and research institutes is being bridged by workshops organized by some private biotechnology companies and institutions. These workshops have been highly effective in our capacity building as they expose our staff and students to the basic practical aspects of biotechnology. The numbers of such workshops are still too few; some of the workshops do not meet up with the required standard, while most of them are still too expensive for the majority of the poor students. Unfortunately, the costs cannot easily be significantly reduced due to the high cost of equipment maintenance, the consumables and basic water and power supply. To ensure that most of the biotechnology students get practical trainings in the various aspects of biotechnology, there is thus a need to subsidize the training costs by the institutions, while efforts are made to provide a more permanent solution by equipping their laboratories, recruiting and training academic staff that will be able to give quality education to our students.

Planning an effective and impactful training workshop will require a detailed organization, focus and creativity. In planning a workshop, the following phase activities are mostly adopted: pre-, during and post-workshop activities. The major pre-workshop activities include setting a defined training workshop goal, appointment of an organizing committee, programme development, budget plan, venue arrangement, publicity, resource persons and workshop manual development. Activities during the workshop are most often a plan for on-venue registration, opening ceremony, the workshop implementation and documentation. Workshop report and networking are the two major post-workshop activities. These three phases are discussed below.

9.5.1 Pre-workshop Activities

This phase of activities is considered as the planning phase, which when successfully planned, the workshop will be said to have attained 50% success. It is akin to the *rate limiting step* in a biological reaction pathway – a step in a metabolic pathway or series of chemical reactions, which determines the overall rate of the other reactions in the pathway. The bulk of the activities to ensure a successful training workshop lie with the following sub-activities that build a strong capacity for any well-organized training workshop (See Figure 9.1).

FIGURE 9.1 Schematic presentation of activities to be carried out at pre-workshop preparation stage.

Defined Training Workshop Goal

Practical biotechnology training should be divided into levels, and each level should be targeted towards mastery of some key biotechnology techniques that form the goal for the particular workshop. This/these goals must be clearly stated which will give a focus to the workshop organizer in coming up with some specific objectives that will actualize the goal. Failure to set a clear goal will amount to waste of time and the entire activity in futility. Target participants are crucial in setting up a training workshop goal, and where necessary, prerequisite techniques and knowledge must be well defined to ensure that all the participants can follow the training. A well-defined training goal will also enable the workshop planners to define scope of the curriculum, identify the appropriate resource persons or specific technical experts required or any special requirement for the training.

Organizing Committee

Second in this phase is the setting up of an organizing committee which will be charged with the responsibility of organizing the entire event. The committee will be saddled with the responsibility of achieving the set objectives so that the overall goal of the workshop can be realized. The organizing committee may have sub-committees to take care of specific tasks with respect to the workshop organization. Such sub-committees may include but not limited to technical, publicity, logistics and entertainment. The overall success of a training workshop hinges on the active role of these sub-committees. Reports from the sub-committees are considered and adopted by the general organizing committee generally referred to as "Local Organizing Committee (LOC)".

Programme/Time Table/Time-Line Development

The organizing body – Department, Institute, Centre, University or Private Company – is responsible for appointing members of the LOC. It is then the responsibility of the LOC to develop the programme for the workshop based on the already defined workshop/training goal. The programme is the planning document that will identify the major workshop activities, assign responsibilities and fix datelines for completing each activity. Assigning specific responsibilities to sub-committees significantly enhances the success of the workshop planning.

Budget

Planning a successful workshop requires a good budget plan. The budget plan translates into how much will be spent and how much will be generated from the workshop (if workshop is not fully funded). The plan will itemize the expenditure that will be incurred and identify the source of fund as income. Expenditure could include, but not limited to, the following:

- Purchase of chemicals, reagents and consumables,
- Equipment hire,
- Provision of tea breaks, lunch (catering),
- Workshop materials (e.g. bags/folders, jotter, pens, name tags, etc.),
- Printing of flyer, poster, banner, etc.,
- Postage of flyer and poster,
- Logistics,
- Payment for resource persons' entitlement (i.e. for travel, accommodation and/or honoraria payments).

The envisaged source of income could include the following:

- Monies that will be generated as registration fees,
- Sponsorship from corporate organizations,
- Grant from the organizing body and others.

Venue

A conducive, equipped and befitting venue is necessary for a successful workshop. Laboratories to be used for hands-on training in Biotechnology should have adequate space and bench surface to enable free movement of participants and provide space for participants to work in small groups. The laboratory should preferably have multiple laboratory wares such as pipettes, glass-wares, centrifuges, vortex mixer, etc. to enable each trainee to participate in all activities. The laboratory should be biosafety certified, depending on the kind of research being conducted, and should also have in place, safety signs and emergency procedures.

All participants would be required to come with their laptops, as most of the workshop information and bioinformatics analysis are web-based. Other facilities that may be required for lecture component of the training include projector, whiteboards, flip charts, chairs, tables, sound equipment, etc. and should be provided by the organizers.

Registration Brochure

A registration brochure should be developed and produced bearing in mind the type of audience that will attend. The brochure should also fit to online registration option. Details to include in the brochure are the theme and aims of the event, programme summary, date of the event, location details, who to attend, registration fee options, closing date for registrations, payment options and contact details of some key members of the organizing committee.

Publicity

There is the need to create wide publicity of the event in the form of advertisement by sending out bulk e-mail to prospective participants, posters and flyers as well as develop a website for the event. It is important to acknowledge participants who register early before the event.

Resource Persons

Resource persons are key to an effective and impactful training workshop. Appointing the right resource persons is a green light to a successful workshop. All the resource persons invited should constantly be in touch, giving them update on the progress of preparations and also obtain assurance from them of their sure attendance. If keynote speakers or other external resource persons will be involved, letters should be briefed early, and their training curriculum and responsibility be clearly defined.

Evaluation Tool

Pre-workshop evaluation is normally required to assess participants' level of knowledge in the planned workshop. For effective training, only candidates with the knowledge/skill should be allowed to attend. At the end of the workshop, a post training evaluation should also be conducted with pertinent questions usually bordering on level of achievement of set out objectives and areas requiring more attention. This will enable the organizers have an effective feedback.

Workshop Manual

Preparation of a workshop manual is an important pre-workshop activity. The manual is a technical guide to the experimental procedures for a hands-on training workshop. A detailed manual will guide participants through reaction steps and highlight the reaction mechanism. Manuals could serve as reference materials to many participants, mostly when they will need to apply the techniques in their laboratories or field work. Protocols should be well described and each step-by-step procedure should be clearly explained to enable participants work with minimal supervision. Manuals should preferably have charts, diagrams, leaflets, pictures or even cartoons to clearly describe each process of the hands-on training. Reference material, websites or important links can be included in manuals.

Equipment and Reagents Pre-test

Pre-test of equipment and reagents is another important pre-workshop activity for a hands-on training workshop in Biotechnology. The pre-test allows the organizers, facilitators and resource persons to ensure the workability of the facilities. The pre-test will test quality of the equipment, quality and sensitivity of reagents and adaptability of the workshop manual.

9.5.2 During the Workshop Activities

Registration and Opening Ceremony

Registration and opening ceremony are the first two events usually undertaken on the first day of most training workshops. A sub-committee will adequately handle a hitch-free on-venue registration and organize a befitting opening ceremony.

Venue Preparation

Preparing the venue can or may include: putting up signage (e.g. room numbers, facility directions), organizing the room(s) (e.g. arrange chairs/tables), making available workshop materials and their distribution. Grouping of participants into small workshop group can be carried out at this stage.

Workshop Implementation

This phase of the organization is the workshop implementation stage in which the training programme comes to life. Having had a very robust and meticulous planning, this phase can be very smooth. Primarily there is the need to establish a learning approach that encourages self-development. A detailed workshop programme of events will provide the necessary guide for a participatory approach to the workshop.

For smooth and steady learning process training, resource persons and facilitators need to organize the sequence of presentations, exercises, reviews and breaks and give appropriate timings in such a way that participants can adequately digest and absorb the workshop contents. The resource persons and facilitators should closely supervise what the participants are doing to ensure that they are doing the right thing. It is very important that the facilitators and resource persons make the information fun and memorable for the participants and get everyone involved.

The following can enhance the learning environment for a more robust workshop success:

- Create more group work but ensure group size is small so that each participant can have the opportunity to be fully involved.
- Find a softer way of correcting error/mistake rather than outright criticism.
- Encourage participants to individually try the techniques.
- Discuss mechanisms of reactions to find alternatives.

Document the Event

Video coverage and photos and use of social media tools are important parts of documentation. Photos and short clip video when posted on website can help promote the activities and may help in future organization of the workshop.

9.5.3 Post-Workshop Activities

Workshop Report

Writing a post-workshop report is an essential component of organizing an effective training workshop. The report helps to document the whole workshop events. The report should capture essentially the organization, finance and participants. Successes, challenges and assessments of the workshop should also be clearly documented (see Figure 9.2).

Networking

The digital age allows a creation of a long-lasting network between participants, resource persons and facilitators. The advantage of the networking is cross-fertilization of ideas and formation of research groups.

Organizing an effective and impactful training workshop is no doubt a herculean task. An effective training workshop is built by following a systematic, step-by step process from planning to execution. Training workshop not carefully planned often fail to meet the intended objectives and participants expectations. However, when carefully planned and executed, training workshops can build capacity, enhance creativity and improve productivity of participants.

9.6 CHALLENGES IN BIOTECHNOLOGY EDUCATION IN NIGERIA

Biotechnology education in Nigeria has been witnessing a steady improvement, and we hope for sustained efforts by all the stakeholders to ensure that Biotechnology plays pivotal roles in the socio-economic development of Nigeria. However, the following issues must be addressed.

9.6.1 Human Resource Development

In principle, it can be said that there is enough (in terms of numbers) human resources for biotechnology education in Nigeria. The core academic staff are available in the several Departments of Biotechnology, as well as scattered across the various traditional

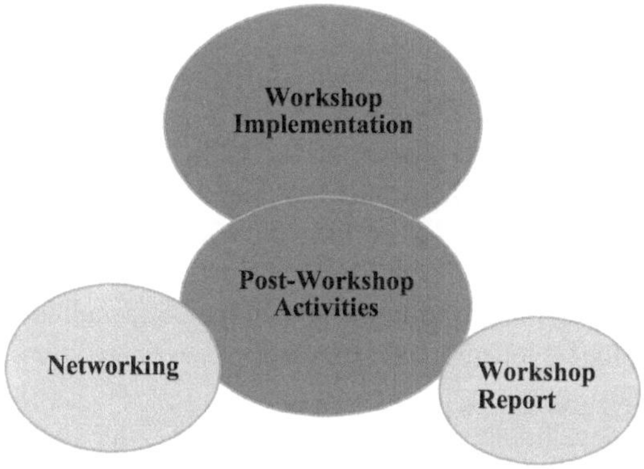

FIGURE 9.2 Schematic presentation of workshop implementation and post-workshop activities.

Departments in the Faculties of Biological Sciences, Agricultural Sciences, Pharmaceutical Sciences, Medical Sciences, Health Science and Technology, Engineering and even Physical Sciences. Unfortunately, most of the staff in the traditional departments lack specialization and in some cases even lack the basic understanding of biotechnology as a discipline. Most of them do not understand the interdisciplinary nature of Biotechnology and tend to be tied up to their traditional departments where they received their training. Furthermore, due partly to the general lack of research infrastructure, many are not engaged in good academic research through which they acquire the knowledge that will guide their teaching. There is therefore a need for continuous training and retraining of the current academic staff in the field of Biotechnology.

Aside from core academic staff, there is also obvious lack of qualified laboratory technologists/personnel to operate and maintain the few equipment in our laboratories. In some departments, many procured equipment remain in cartons because nobody knows how to operate them. In other cases, such equipment are used for just a short period, and once they develop even the least fault, they are abandoned because nobody knows how to repair. Asking the manufacturers to come and repair such equipment is not an option because of the distance and financial implications.

9.6.2 Infrastructural Deficiencies

Biotechnology education is expensive because most of the equipment that are required for hands-on training, research and development are very expensive. Most of the universities offering Biotechnology programmes currently lack most of the basic laboratory equipment and the few available ones are not properly maintained. Even where some of these equipment exist, they are scattered in various departments and in most cases not readily available to the students. Nevertheless, some universities have fairly well-equipped laboratories and some have central laboratories that are jointly used by the various departments in the university. In the current financial difficulty in Nigeria, the central laboratory model is the best since it allows for optimum utilization of the limited resources. There should also be well-trained laboratory technologists and technical officers for proper maintenance of the equipment.

9.7 CONCLUSIONS AND RECOMMENDATIONS

In order to ensure that Nigeria does not continue to lag behind in Biotechnology, government incentives are needed enormously to play an overriding role in driving the direction of biotechnology education in the country. Government should encourage and fund biotechnology education at all the levels but especially at postgraduate level. As part of this goal, governments at both the national and state levels should provide cornerstone funding for establishment of bio-enterprises, state-sponsored programmes and the development of core research capabilities at our universities, as it is done in most of the rapidly developing economies in Asia, in order to attract bright inventive minds from the postgraduate schools of our universities to bioscience entrepreneurship. There is no doubt that locating such projects in our university campuses would boost expertise and employment. In many developed countries, for example, government-sponsored programmes offer large incentives to create biotechnology infrastructure and develop biotechnology and life science parks, which help to recruit local and foreign talents to the country. Perhaps more than at any time in history, Nigeria presently needs the best cutting-edge technologies that could be adopted to bear on the myriad of problems facing economic growth and development of the country. Biotechnology should be considered as a strategic emerging industry and plans should be afoot to prioritize the development of this cutting-edge technology in our university campuses to enhance entrepreneurial strength that would capitalize on the huge potential that presently lies underexploited in the country. There is need to also strengthen university–industry partnership as a boost to the biotechnology landscape in Nigeria.

REFERENCES

120 Universities that offers Biotechnology in Nigeria. Website: https://fatherprada.com/the-list/119-list-of-nigeria-universities-offering-biotechnology/ (last updated 29th June, 2018; accessed 29/03/2020).

2020/21 JAMB Brochure Online. Website: http://ibass.jamb.gov.ng/assets/uploads/2018-09-03-221207-sciences.pdf (accessed 30/03/2020).

Nigerian Universities Commission (NUC) (2018) Biotechnologist Propose New Curriculum. Website: https://www.nuc.edu.ng/biotechnologists-propose-new-curriculum/ (accessed 26/03/2020).

PLA Notes (1999) A Brief Guide to Organising Workshops for Training, Orientation and Exposure in Participatory Approaches. Issue 35, pp. 92–96, IIED London.

Toscany Academy (2012) The Fate of Biotechnology Program in Nigeria, Nov. 16th, 2012. Website: http://toscanyacademy.com/blog/courses-and-programs/the-fate-of-biotechnology-program-in-nigeria/%20 (accessed 27/03/2020).

10 Challenge of Science, Technology, Engineering, and Mathematics (STEM) in Technological Development of Nigeria

Olu Odeyemi
Obafemi Awolowo University

Modupe Stella Ayilara
North-West University

CONTENTS

DOI: 10.1201/9781003179177-10

10.1 INTRODUCTION

The goal of Nigeria vision 20:20/20 is to make Nigeria one of the 20 largest economies in the world, able to consolidate its leadership role in Africa and establish itself as a significant player in the global economic and political arena (Jatto, 2010; Saidu, 2011; Steinfioff and Burgess, 1993). However, it is a shame that with her many endowed natural and human resources, Nigeria is still behind in areas such as biotechnology, electronics, genomics, information and communication technology, food technology, materials technology and production and process technology (Uwaifo and Uddin, 2009). We still depend on foreign products and even go for foreign expatriates for operations and maintenance of many of our machines when we have many trained engineers around us. The actualization of democracy in 1999 made many people believe that Nigeria was bound to experience developmental changes in all sectors. Nearly two decades after transition from military regime to democratic regime, Nigeria is still poor scientifically and technologically. Today, political changes provide the nation with the opportunity for a fundamental review of policies and social programmes that will prepare the country to meet the challenges of the new century (Siyanbola et al., 2016). For example, Vision 20:20:20 envisages the country to be among the first 20 countries in terms of industrialization, democracy, educated human resources etc. (Olaniyan and Adetoye, 2016). By year 2020, with the Millennium Development Goals (MDG5) which include policies to eradicate extreme poverty, achieving Universal Basic Education, promoting gender equality and empowerment of women, reducing child mortality and improving maternal health, controlling of HIV/AIDS, malaria and other deadly diseases, developing global partnership with other countries and ensuring environmental sustainability (Organization, 2015). These policies all aimed at giving special attention to core economic, political and social variables to speed up development of Nigeria as a nation. The objective of this write up is to evaluate the challenges of science, technology, engineering and mathematics in Nigeria. This chapter addresses the challenges of Science, Technology, Engineering and Mathematics in the technological development of Nigeria.

10.2 EVIDENCE OF SCIENTIFIC AND TECHNOLOGICAL BACKWARDNESS IN NIGERIA

Nigeria's backwardness in the aspects of science and technology is obvious, and this can be emphasized by the:

a. inability to produce capital goods such as cars, buses, tractors, lathe machines, drilling machines, trains and other earth moving equipment.
b. inability to exploit the nation's prodigious natural resources except with the help of foreigners who will provide the technology and expertise required for such exploitation.
c. dependence on advanced countries for supply of spare parts for industrial machinery.

d. inability to produce military hardware and software needed for national defence and security.
e. export of raw materials such as crude oil to other countries as against finished products such as premium motor spirit (gasoline), kerosene, aviation fuel and diesel.
f. inability to mechanize agriculture since crude implements such as cutlasses and hoes are still used for agricultural production activities by a large percentage of the farmers.
g. dependence almost entirely on rain-fed agriculture with little or no irrigation to ensure food production year round.
h. inability to preserve and process food crops in an effort to eliminate the annual 40% loss of harvest to pests and diseases.
i. inability to guarantee food security for citizenry as only about 5% of the people can afford the required 2,350 calories of food per person per day with the rest making to do with about 500 calories per person per day. Hence, Nigeria depends on food importation to feed its teeming population which now stands at 180 million people.
j. inability to power economic development of the country through electricity generation and distribution as it requires at least 40,000 megawatts but currently generates only 4,000 megawatts.
k. inability to maintain all basic and essential infrastructural facilities. For instance, according the former Corps Marshal and Chief Executive of the Federal Road Safety Commission (FRSC), Mr. Osita Chidoka, at the 2011 Nigerian Institute of Safety Professional's Conference in Warri, "Nigeria is currently 191st out of 192 countries of the world's unsafe roads."
l. inability to provide gainful employment for its over 40 million secondary school leavers and university graduates.
m. inability to educate over 10 million children of school age who presently are not in school.
n. inability to embark on urgent training and massive production of Nigerians in Science, Technology, Engineering and Mathematics (STEM) to ensure future technological and economic development.
o. poorly equipped, inadequately staffed and underfunded educational institutions at all levels of education.
p. inability to collate and commercialize research findings.

10.3 ENTREPRENEURSHIP

There are different definitions of entrepreneurship. For instance, Sobel (2008) explained entrepreneurship as the ability to develop a new venture or apply a new approach to an old business, while Salami (2011) defined entrepreneurship as the ability to seek investment opportunity for exploitation. On the other hand, entrepreneurship has been linked to the creation of five basic "new combinations" of (a) introduction of a new product, (b) a new method of production, (c) opening a new market, (d) conquest of a new source of supply, and (e) creating a new organization.

Generally speaking, entrepreneurs can be defined as innovators, creators, coordinators, initiators, organizers, gap-fillers, leaders and risk takers.

Entrepreneurship may be classified into two broad types, viz., opportunity-based entrepreneurship and necessity-based entrepreneurship. Opportunity-based entrepreneurs perceive a business opportunity and decide to pursue it. For such people, a period of economic meltdown is a perfect time to pursue new market opportunities. That is, recession tends to favour the creative and innovative entrepreneurs. On the other hand, a necessity-based entrepreneurship is borne out of a survival strategy when there are no other options to earn a living and by compulsion, entrepreneurship is chosen as a career. For instance, during the Nigerian Civil War of 1967–1970, the secessionist Biafrans were forced to device various innovative survival strategies to meet their daily needs for food, fuel, transportation, weapons, etc.

10.3.1 Entrepreneurship Strategies for Enhancing Food Production and Economic Transformation in Nigeria

Agriculture is absolutely essential to the transformation of the Nigerian economy as it is the largest employer of the workforce (70%) and the biggest contributor to the country's Gross Domestic Product (42%). Farmers grow the crops needed to feed, clothe and produce energy for the country. Hence, by investing in farmers, the country can significantly improve its economic status (Nwangwu, 2019). Some of the entrepreneurial strategies for achieving food security and economic transformation of Nigeria include the following:

i. **Increasing land area under cultivation:**

The Nigerian agricultural sector has the potential to grow its current output, valued at N 15.25 trillion, by 160% to N 39.4 trillion by 2030. This growth potential could be achieved by increasing both the productivity and total acreage of cultivated land since about 38% of Nigeria's arable land of 80 million hectares is yet to be tapped. The area under cultivation can be increased through sustainable utilization of forested land, use of alley cropping and application of organic fertilizers to rehabilitate eroded and denuded soils. Agricultural productivity can also be enhanced by irrigation of arid soils and deserts through water harvesting and reclamation of denuded and hydromorphic lands. Scientifically, food is free because it is a product of photosynthetic activity of green plants that capture free energy of the sun and transform same into grains of rice, wheat, maize and tubers like yam, cassava, potato, etc. It is therefore a 'sin' of illiteracy, laziness, ignorance, lack of technical know-how and lack of governmental political will for Nigeria to continue to suffer food insecurity and experience poverty in the midst of almost 40% of her arable land 'begging' for cultivation. Both opportunity-based and necessity-based entrepreneurs should be encouraged by the three tiers of government in Nigeria to exploit the vast untapped agricultural lands for agrarian and economic transformation of the country (Figure 10.1).

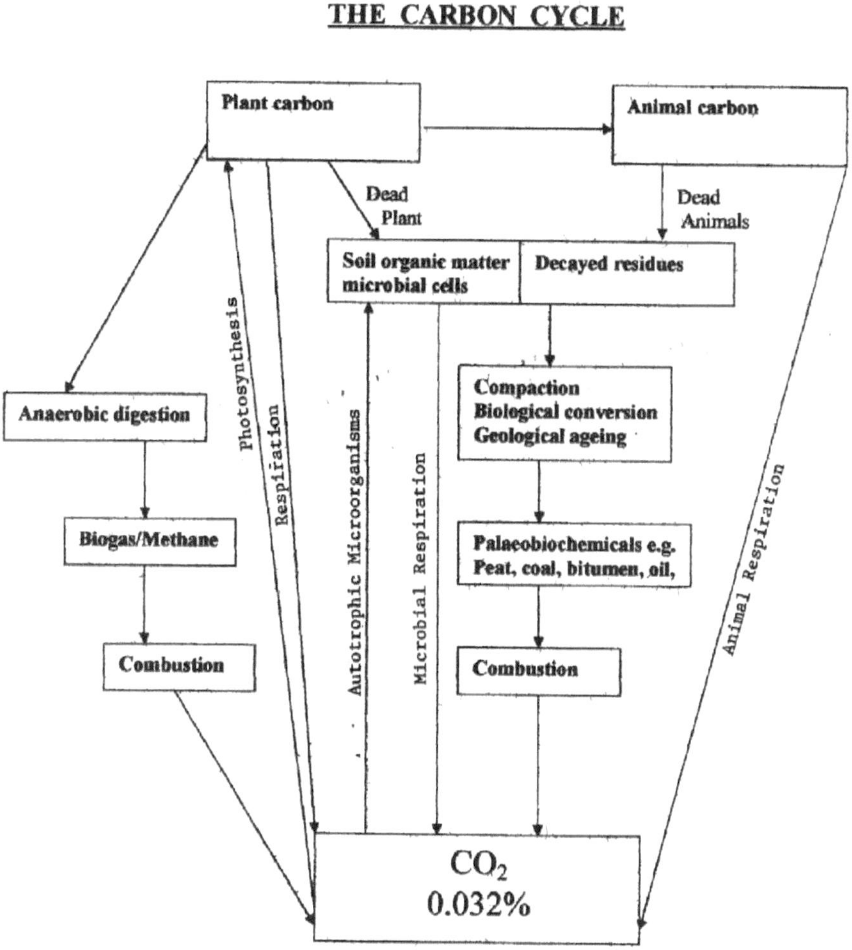

FIGURE 10.1 The carbon cycle.

ii. **Enhancing quantity and quality of food production per unit area of land:**

Worldwide, agriculture accounts for 34% of the land area and 70% of water consumption. A variety of practices such as conservation agriculture and technologies such as drip irrigation or improved rainwater harvesting can help manage natural resources more sustainably. Since land area is limited, there must be continuous search and research for production needs to increase the quantity and quality of food per unit area of land. This objective can be achieved through some novel agricultural practices such as the use of biological nitrogen fixation like the *ENPOST Rhizobium Inoculant* developed by one of the authors (Odeyemi, 2008). Other beneficial practices include mycorrhizal phosphate nutrition, multiple cropping, relay cropping, intercropping and use of improved crop varieties and breeds of animals. Proper consideration must also be given to more efficient use of fertilizers and soil conditioners as well as improved soil and water management, weed control and pest management. For instance, the FAO (2004) employed our ENPOST Rhizobium biofertilizer to enhance pea yields of cowpea by about 50% in Niger, Nasarawa, Katsina, Bauchi and Kogi states of Nigeria (Odeyemi, 1991). By increasing crop yields, natural habitats can be protected from being cultivated, and forests and bushes would continue to sequester carbon to mitigate emissions promoting climate change. For instance, the average yield of rice in Nigeria is about 5 t/ha, but through the use of improved varieties and good agricultural practices, the yield can be boosted to about 10 t/ha. Local production of rice should be encouraged urgently in order to reduce the current alarming volume of rice importation which stands at one billion Naira per month.

iii. **Technology innovation:**

Most of Nigerian farmers find it very difficult to adapt to new innovations partly due to illiteracy and lack of effective training and extension services (Uguru et al., 2015). Yet, strengthening rural innovation systems, developing effective clusters that can add value to unprocessed raw materials and promoting value chains across such diverse sectors as horticulture, food processing and packaging, food storage and transportation, food safety, distribution systems and exports are all central to moving beyond subsistence agriculture to generate growth, spawn economic transformation and engender prosperity.

Agricultural productivity, entrepreneurship and value addition foster economic growth especially in a rural-based economy like that of Nigeria. Unfortunately, the small-scale farmers and research institutes do not interact in ways that can accelerate the transition from subsistence to commercial value-added agriculture. Therefore, a comprehensive approach to agricultural development in Nigeria must involve the creation of an innovation system that fosters interactions among government, industry, farmers, academia and civil society, all of which are essential components of agricultural transformation. In particular, efforts should be made to bring higher technical education such as available at the Federal Colleges of Agriculture, Agricultural Research Institutes and the Universities to the service of agriculture and the Nigerian economy. Agriculture must be viewed as a knowledge-based entrepreneurial activity; hence, the Nigerian government must invest adequately in agricultural research as part of a larger agenda to promote innovation, build human capacity and stimulate entrepreneurship.

iv. **More food from livestock and fishery:**

Animal production can be enhanced through genetic improvement, adaptation to environmental stresses, resistance to pests, parasites and diseases and improved feeding and nutrition. Efficient management and exploitation of aquatic species in the numerous bodies of water in Nigeria can ensure increased production of fish and other aquatic resources. Formulation of cheap local fish feeds and other animal feeds is important in reducing cost of producing animal proteins (Bhosale et al., 2010).

Improved processing, curing, packaging and preservation of animal products are also relevant. Aggressive marketing strategies at local, national and international arena are also necessary for improved production of animal products.

v. **Improvement of post-harvest technology:**

In Nigeria, about 40% of crop harvest is lost annually to pests and diseases (Heeb et al., 2019). Most of the approximately 100,000 known species of fungi are strictly saprophytic, living on dead organic matter that they help decompose and about 100 different species cause diseases of human and animals, but more than 8,000 species cause plant diseases, especially post-harvest diseases, which are very damaging. For instance, the rust diseases caused by different Basidiomycetes are among the most destructive diseases known to humanity. They have caused numerous famines and economic depressions and still reduce the world grain harvest by an estimated 10% per year. Improved techniques to reduce losses may include better timing of harvests, improved handling, threshing, transportation, drying, curing and conditioning of farm produce. Effective control of insects, rodents, bacteria, mites, termites, moisture, etc. in stores, bins and warehouses is also crucial (Fried et al., 2017). Improved milling, processing, packaging, preservation and fortification are essential post-harvest technologies. Thus, by investing in storage and processing facilities to reduce post-harvest losses and transport links to access markets, farmers can more reliably increase the quantity and quality of crops they grow and bring to markets (Henz, 2017).

vi. **Improvement of infrastructure and marketing:**

To stimulate entrepreneurial interest in agriculture, the government must invest in enabling infrastructure such as roads, railways, electricity, fertilizer factories, seed banks, human and animal clinics, treated water supplies and information and telecommunication networks (Daud et al., 2018). It is also necessary to review the 1999 constitution of Nigeria to liberalize the accessibility to and ownership of arable lands by farmers and youths. Sustainable and profitable agriculture can only thrive when there are reliable, viable and functional markets for farm produce. The local, state and federal governments must create community, intra-state, inter-state, national, regional and international market accessibilities for our farmers. For instance, Nigeria should intensify fostering of regional markets among ECOWAS countries and learn from the achievements of Rwanda and Malawi or those of China, Brazil and India, in this regard. Indeed, there is the need for countries in central, eastern, western, southern and northern Africa to work together cooperatively at the regional level to scale up agricultural development, marketing and economic transformation. South-south and north-south cooperation is also critical to creating agricultural markets and fast-tracking economic transformation of Africa in general and Nigeria in particular.

vii. **Financing of entrepreneurship in agriculture:**

It is well known that economic returns on agricultural investments are very low; hence, banks are always reluctant to advance loans for agricultural enterprises which they perceive as extremely high risk, unpredictable and unreliable endeavours (Osabohien et al., 2018). Yet, if Nigerian agriculture must transit from the subsistence level to the entrepreneurial and commercial status, which can guarantee profitability, it must enjoy private- and public-sector funding.

Several efforts have been made by previous governments in Nigeria to support agricultural financing but not much success has been achieved in this regard. However, recently, the Central Bank of Nigeria (CBN) launched the N 77.5 billion Nigeria Incentive-based Risk Sharing System of Agricultural Lending (NIRSAL) meant to de-risk lending to the sector. According to the CBN, NIRSAL is aimed at lifting Nigerian agriculture from the subsistence to the entrepreneurial and commercial level to engender profitability in agricultural enterprises. NIRSAL is also expected to help banks understand profitable agricultural financing. The five components of NIRSAL are: (a) Risk sharing facilities which are allocated N 46.5 billion; (b) Insurance facilities, N 4.65 billion; (c) Technical assistance facilities, N9.3 billion; (d) Agric bank rating system, N1.55 billion; and (e) Bank incentive mechanism, N 15.5 billion. As laudable as the NIRSAL proposal appears, the CBN still needs to sensitize and educate all the stakeholders about the workability of the scheme. For example, how will the federal and state governments collaborate with the CBN in implementing the scheme? How will the thousands of peasant farmers and prospectus agricultural entrepreneurs access the loan without much stress?

10.4 IMPORTANCE OF SCIENCE, TECHNOLOGY, ENGINEERING AND MATHEMATICS (STEM)

Sweet as the goal of Nigerian's vision 20:20:20 may sound; every right minded person in this nation will agree with me that without science and technology, the attainment of the core values of vision 20:20:20 will be elusive. It will definitely be impossible to achieve vision 20:20:20 without successfully harnessing and developing technological manpower. In fact no nation can achieve developmental growth without technology. Technology is defined as the know-how and creative processes that may assist people to utilize tools, resources and systems to solve problems and to enhance control over the natural and man-made environment. Its importance in improving human conditions and the nation's development cannot be overemphasized. This can be by ideas and taking action, as well as developing techniques and products that satisfy human needs.

To actualize the goal of vision 20:20:20, Nigeria must adequately embark on training her citizenry on the need to be technologically literate, leading to self-reliance and sustainability. Effective technology programmes encourage students to be productive, innovative and enterprising. In fact, technological education should be a compulsory learning area in our national curriculum for all students from secondary schools and then provide pathways for them to access technological opportunities beyond secondary education. In our contemporary world, only technologically inclined nations command the skills necessary for sustainable economic growth and a better quality of life for her citizenry. Technology helps in building a nation effectively and efficiently and also helps in opening business relationship with other friendly nations. Over the years, our inability to efficiently equip our youth with sound science and technology through technological education has put us at a low level in technology and has perpetually made Nigeria a developing nation. In fact, the difference between developed, developing and underdeveloped countries rests on the ability of the developed countries to convert scientific ideas to usable technology while the developing and underdeveloped countries are yet to effectively do so. For a vibrant technology transfer, it is imperative to first liberate Nigeria from technological illiteracy through technological education. By definition, technological education is the training of technical personnel for the purposes of initiating, facilitating and implementing the technological literacy to the young minds. The objective of technological education is to upgrade students' level of technological literacy. Technological education focuses on increasing students' technological literacy through the enhancement of their technological practice across technological areas and contexts. Technology education needs to focus on enhancing students' technological practice (Serdyukov, 2017). This will allow them to become empowered decision makers, participating as mutually constitutive members within their learning community and those communities within which they may participate in the future.

10.5 IMPORTANCE OF TECHNOLOGICAL EDUCATION

Technological education plays a vital role in human resource development of the country by creating skilled manpower, enhancing industrial productivity and improving the quality of life. Technological education covers courses and programmes in engineering, science, technology, biotechnology, management, architecture, town planning, pharmacy, applied arts and crafts, hotel management and catering technology. To respond to the recurrent and emerging economic and social needs of our country our youths must be made to acquire the understanding of the role of science and technology in society. Youths must also acquire skills of analysis and problem-solving as well as scientific and technological skills which will allow them maximum flexibility and adaptability in their future employment and other aspects of life. Over the years, we have failed as a nation to empower our youths with technological and scientific career. The consequence of this failure is turning out of schools, restless and disconnected generation of youths. Today, the nation faces crisis of scarcity of scientific and technological manpower. To achieve technological and economic development in the year 2020 and beyond, Nigeria needs to sponsor many students abroad to acquire necessary expertise in various critical areas of science, engineering and technology.

10.6 SOME OF THE CAUSES OF TECHNOLOGICAL BACKWARDNESS IN NIGERIA

10.6.1 Discouragement of Technological Growth by Our Colonial Masters

The philosophy of Nigerian education during the colonial period was built on wrong policy as evidenced by the following statement of Lord Lugard (1921): "The chief function of government primary and secondary schools among primitive communities is to train the more promising boys from the village schools as teachers for the schools, as clerks for the local native courts, and as interpreters" (I. O. Osokoya; <Osokoya. pdf (globalacademicgroup.com)>). There were many reasons why the British came to Nigeria. One of such reasons was to spread the Christian religious faith as missionaries (Ndubuisi, 2018), but the principal reason was economic. Nigeria was perceived as a ready market for British alcoholic spirits, dane guns, gun powder, mirrors and slaves.

It should be noted that before the advent of colonialism, Nigerians were weaving their own clothes, smelting bronze, tanning hides and skin to produce shoes and bags, fabricating hoes, cutlasses, bows, arrows and casting an object as intricate as the "Benin/Festac Mask" which was among the numerous art objects stolen by the British. The colonialists discouraged further development of Nigerian technology as they reasoned it was a threat to the marketing of goods imported from Europe (Isioto et al., 2017). For instance, "ogogoro" was declared an illicit gin by the colonialists, and whoever was caught producing, marketing or drinking it was prosecuted. However, in the 1960s, the late Dr. Tai Solarin and Prof. Femi Odeyemi of Chemistry Department, University of Lagos (one of the authors (Prof. Olu Odeyemi) older brother), fought vehemently to legalize the locally brewed gin (ogogoro). This legalization led to the establishment of intercontinental distillers (industrial alcoholic spirit producers) by Prof. Femi Odeyemi and Co. at Ota, Ogun State.

10.6.2 Weak Industrial Policies after Independence

The various industrial policies put in place by Nigerian government after attainment of independence were patently weak and lacked comprehensiveness. The major thrust of the policies were (1) building of assembly plants such as VW and Peugeot vehicles in Nigeria, (2) importation of Completely Knocked Down (CKD) parts into the country for use by the assembly plants, (3) establishment of steel plants such as Delta Steel Plant and Ajaokuta Steel Plant and associated foundries to produce automobile parts and raw materials for steel and other fabrication activities and (4) establishment of machine tool companies (e.g. Osogbo Machine Tool company) to produce capital goods. Unfortunately, this import substitution industrial strategy did not go beyond the stage of building the assembly plants, as the technical partners know that if Nigeria stops importing the CKD parts, their companies in Europe might stop production and fold up. In any case, almost all the steel companies, the assembly plants and the machine tool companies are now in a state of disrepair due to abject neglect and complete lack of maintenance.

10.6.3 Poorly Equipped Educational Institutions

Almost all educational institutions from primary to secondary and tertiary levels in Nigeria are poorly equipped, understaffed and underfunded (Amaechi et al., 2018). In particular, our Universities, Polytechnics and Colleges of Technology that are supposed to produce competent and proficient engineers, scientists, agriculturists and technologists are completely lacking in state-of-the-art equipment and qualified, adequate and competitive academic staff. Hence, the products graduating from these substandard institutions are obviously poorly baked and consequently incapable of outstanding contribution to the country's technological development. Apart from the developed countries, the so-called BRICK countries, i.e. Brazil, Russia, India, China and Korea, have developed well-funded, properly staffed and adequately equipped Universities and technological institutions to produce high-quality graduates in fields of instructions such as information technology, electronics, computer engineering, biotechnology, space science, genetic engineering, etc. to fast-track industrial and economic development. No wonder, the economies of these countries compare favourably with those of the countries of Europe and North America. The inadequacy of teaching, laboratory, research and workshop facilities has contributed to the diminution of the quality of graduates, especially science and engineering graduates in Nigeria. Students can be categorized into three groups, namely, verbalizers, visualizers and doers (Uwaifo, 2010). The verbalizers are those who learn easily if information is presented in written or spoken form. They benefit maximally from lectures, tutorials and hand-outs. Visualizers learn easily when information is presented in pictorial or diagrammatic form, while the doers learn more readily when information is presented by practical demonstrations by the lecturer. The inadequacy of basic facilities would put the visualizers and the doers at a disadvantage. The visualizers may also have problems in a class with a large student population. The overall effect of this entire scenario is that only a small percentage of our student population benefit from the present pedagogical system. Additionally, accessibility to internet connectivity has been hampered by power fluctuations and inadequate bandwidth. All these problems must be rectified in order for Nigeria institutions to produce high-calibre graduates required to power technological development.

10.7 STRATEGIES FOR ENHANCING TECHNOLOGICAL EDUCATION FOR NATIONAL TRANSFORMATION

10.7.1 Eradication of Extreme Poverty and Hunger

This is GOAL 1 of the UN's MDGs which if properly executed holds the promise of saving of lives; empowering women; addressing the scourge of illiteracy, hunger and malnutrition; and ensuring that Nigerian children have access to education of good quality and good health to lead productive lives. Developmental goals can be unattainable without quantitative and qualitative education especially in Nigeria with a high percentage of the population living below poverty level (Ajodo-Adebanjoko and Walter, 2014). An English adage says that a hungry man is an angry man, and a Yoruba adage says "ebikiiwonukioranmiran wo", meaning when you are hungry, nothing else counts or matters. Nigeria must therefore invest a high percentage of its petro-naira into agriculture in order to be able to feed its population of about 180 million. Nigeria has about 80 million hectares of arable land but only about half of it is under cultivation. The Land Use Act of 1999 should be repealed to liberalize ownership or accessibility to land by intending farmers especially among the jobless youths who should be enticed to take up agriculture. Food security entails adequate food availability, accessibility and utilization. According to the International Food Policy Research Institute (IFPRI), the minimum food needed for a healthy and active life is 2,350 calories per person per day. Most Nigerians can hardly afford 1,000 calories per capita per day. Food insecurity and malnutrition are unlikely to disappear in Nigeria in the next 10 years. However, broad-based agricultural and rural development plus greater public investment in agricultural research and better management of natural resources will be required to eliminate food insecurity in Nigeria. A well-nourished and highly educated citizenry will unleash their heretofore latent abilities to catalyse technological development and economic transformation of the nation.

10.7.2 Greater Emphasis on Technological Education

Governments at all levels must vote and spend more funds on education in order to arrest the shameful rot in our educational system, and they should stop paying lip service to raising the standard of education especially through their talk shows and wasteful rituals of education summits. In particular, training of technical and technological manpower must be given a top priority. Most of the problems militating against production of high-quality scientists and engineers must be squarely addressed. Such encumbrances include poor funding, inadequate facilities, non-availability of adequate human capacity, brain drain, poor staff training and retention profiles. Other factors include poorly equipped laboratories, defective and substandard curricula, weak University and Polytechnic industry partnership, archaic traditional approach to teaching, inadequate infrastructural facilities such as electricity, water and classrooms as well as lack of quality control mechanisms.

10.7.3 Copying of Technological Items Already in the Market

This proposal requires laboratories, workshops and other necessary facilities be developed for component analysis and for building prototypes of items to be produced (Collins, 2015). The idea is to knock down products of interest in the workshops, study and analyse each component in the laboratories to ascertain chemical composition, physical properties and other production parameters of interest and replicate such items for future mass production. Japan employed this method of technology transfer in the 1970s and 1980s to achieve industrial and economic growth. Nigerian government should encourage the so-called "igbo made" goods by improving their quality to meet those of imported items.

10.7.4 Industrial Espionage

Highly technical, nuclear and defence technologies are closely guarded by their proprietors. Such secrets can only be obtained either by direct heavy financial investments or through espionage (Kennedy and Lim, 2018). Spies are often employed to collect such top secrets and company documents required for developing such products, which they pass on to their sponsors for free. Many countries around the world engage in this type of industrial espionage especially through their embassies and science and military attachés. Nigeria can also benefit from this well-known but seldom discussed practice to fast-track its technological development and economic growth.

10.7.5 Provision and Maintenance of Infrastructural Facilities

Nigeria can take a bold step towards becoming a technologically advanced country by simply providing and maintaining basic infrastructural facilities (Compagnucci and Spigarelli, 2018). These facilities include such as road, railways, electricity, oil refineries, treated water supplies, fertilizer factories, airports, seaports, information and telecommunication networks, educational institutions, inland waterways, stadia, museums, tourism facilities, textile factories, paper mills, aluminium smelting factories, iron and steel production factories. One of the authors (Prof. Olu Odeyemi) wrote letters to President Olusegun Obasanjo in 1999, urging him to establish a Ministry of Infrastructural Maintenance probably as an offshoot of the Ministry of Works, in order to achieve a continuous and sustainable maintenance of our massive infrastructural facilities. Subsequently, the Federal Roads Maintenance Agency (FERMA) was established in July 2003, though the agency appears not effective, probably due to poor funding. The poor state of Nigerian roads is a major aspect of the nation's infrastructural deficit and is a significant contributor to the unbearably high cost of doing business in Nigeria. If Nigeria has about 30 million jobless youths, why can't we deploy them to maintain our sadly deplorable roads on local government by local government basis at minimal stipends?

10.7.6 Bold and Ambitious Energy Provision

Apart from agricultural revolution, an abundant provision of energy is a key ingredient of industrial evolution (Jin et al., 2018). These are major factors that launched Europe and America into the technological age. As Nigeria is also desirous of becoming a technologically advanced country, bold and ambitious provision of energy is a sine-qua-non for attaining such a vision. Fortunately, Nigeria is blessed naturally with vast deposits of palaeobiochemicals such as peat, coal, bitumen, crude oil and gas. These natural endowments can be processed to generate energy needed for industrial development. Nuclear technology can also be employed to generate electricity via processed uranium (which is bountiful in Nigeria) as it is done in most of the developed countries of the world.

Electricity for powering technological transformation can also be obtained through renewable energy sources such as solar energy, biomass energy, wind energy and geothermal energy.

With Nigeria's proven oil reserves of 35.2 billion barrels, gas reserves of 176 trillion cubic feet and coal reserves of about 3 billion metric tons, we have no reason to continually be in the dark. At all cost, Nigeria must acquire the necessary know-how to transform these abundant resources into energy for lighting and national development. The Power Holding Company of Nigeria (PHCN) as presently constituted can never light up or power the country sustainably. As soon as feasible, the PHCN must be completely re-privatized to enhance its efficiency in an effort to achieve the vision 20:20:20.

10.7.7 Wealth Creation from Nigeria Mineral Resources

Nigeria is an immensely mineral rich country liberally endowed as can be seen in Table 10.1. Nigeria is reputed to be one of the few countries of the world that is endowed with virtually all the minerals in the Chemistry's Periodic Table. By

TABLE 10.1
Location, Reserves and Some Uses of the Solid Minerals Found in Nigeria (Odeyemi, 2014)

Mineral	Location	Reserves	Some Uses
Tantalite	Cross River, Ekiti, Kogi, Kwara, Nasarawa	No data	For making capacitors, resistors
Kaolin	Akwa Ibom, Anambra, Bauchi, Bayelsa, Ekiti, Imo, Katsina, Kebbi, Kogi, Ogun, Ondo, Plateau, Rivers	3 billion tons	For making ceramics, toothpaste, in light bulbs, cosmetics, paints, etc.
Mica	Ekiti, Kogi, Kwara, Nasarawa, Oyo	No data	For capacitors, paints, well-drilling, inert filler and mould release in plastics industry, for decorative coatings, tiles, etc.
Baryte	Benue, Cross River, Nasarawa, Plateau, Taraba, Zamfara	7.5 million tons	Drilling fluids in oil and gas exploration, as filling in paints and plastics, friction products for automobiles, etc.
Coal & Lignite	Enugu, Abia, Adamawa, Anambra, Bauchi, Ondo, Benue, Cross River, Delta, Ebonyi, Edo Gombe, Imo, Kogi, Nasarawa, Plateau	3 billion ton	Coal is used for electricity generation, cement manufacture, industrial process heating.
Rutile	Bauchi, Cross River, Kaduna, Plateau	No data	Used in paint, plastics and paper industry. It is also used in electrode coverings and sunscreens.
Talc	Ekiti, Kaduna, Kogi, Niger, Osun, Ogun	Over 100 million tons	Widely known as talcum powder
Bismuth	Kaduna	No data	Pigment in eye shadows, hair sprays and nail polishes and as an ingredient in some pharmaceuticals
Gypsum	Adamawa, Edo, Gombe, Ogun, Sokoto, Yobe	Over 1 billion tons	Manufacture of cement, plaster of paris (P.O.P), classroom chalk
Marble	Edo, FCT, Kogi, Kwara, Nasarawa, Oyo	No data	For sculpture and as building materials
Gemstones	Bauchi, Kaduna, Kogi, Kwara, Nasarawa, Niger, Ogun, Oyo, Plateau, Taraba	No data	Used as ornamental stones, making jewelleries, etc.
Feldspar	Bauchi, Borno, FCT, Kaduna, Kogi	No data	Used in making glass, ceramics, K-Ar dating, etc.
Gold	FCT, Kaduna, Kano, Katsina, Kebbi, Kogi, Kwara, Niger, Osun, Zamfara	No data	For making jewelleries, for coinage, in dentistry, electronics, etc.
Clay	In all the States of the Federation	No data	For sculpture, ceramics material, for making earthenware and porcelain, etc.

(*Continued*)

TABLE 10.1 (*Continued*)
Location, Reserves and Some Uses of the Solid Minerals Found in Nigeria (Odeyemi, 2014)

Mineral	Location	Reserves	Some Uses
Silver	Ebonyi, Kano	No data	For making ornaments, jewellery, high-value tableware, utensils, currency coins, conductors, etc.
Ilmenite	Benue, Cross River, Kaduna, Plateau	No data	In titanium dioxide production
Limestone	Benue, Cross River, Ebonyi, Edo, Gombe, Kogi, Ogun, Sokoto	2.3 trillion tons	Building industry, production of quicklime, slake lime, cement, mortar, glass making, and collectives used as solid base for road construction
Columbite	Bauchi, Cross River, Kaduna, Kano, Kwara, Nasarawa, Plateau	No data	Source of niobium used as an alloy of steel.
Cassiterite	Bauchi, Cross -River, Kaduna, Kano, Kwara, Nasarawa, Plateau	No data	A source of tin, also used as a gem stone.
Diatomite	Borno, Yobe	Over 200,000 tons	Used as filtration material, mild abrasive, absorbent for liquids, thermal insulator, etc.
Phosphate	Ogun, Sokoto	No data	In phosphate fertilizer
Manganese	Katsina, Kebbi, Zamfara	Over 4 million tons	For making alloy, in manufacture of battery, to boost octane rating in gasoline, etc.
Silica Sand	Delta, Jigawa, Kano, Lagos, Ondo, Rivers	No data	Manufacture of glass and bottle, optical fibre for telecommunications, ceramics, etc.
Flourite	Bauchi, Ebonyi, Plateau, Taraba	No data	For making opalescent glasses and enamels, telescopes and camera lens, cooking utensils, beads in jewellery, etc.
Bitumen	Edo, Lagos, Ondo, Ogun	About 42 billion tons	To pave roads, building industry, production of quicklime, slake lime, cement, mortar, solid base for road construction, glass making
Lead	Cross River, Ebonyi, FCT, Plateau, Zamfara	10 million tons	Used for piping, roofing materials, ammunition, toys, ornaments, coffins and protective coating.
Zinc	Cross River, Ebonyi, FCT, Plateau, Zamfara	10 million tons	Used for protective coating for iron and steel, making brass, soldering fluid, screens for television tubes and fluorescent coatings.

(*Continued*)

TABLE 10.1 (*Continued*)
Location, Reserves and Some Uses of the Solid Minerals Found in Nigeria (Odeyemi, 2014)

Mineral	Location	Reserves	Some Uses
Bentonite	Borno, Edo, Kogi, Ogun, Ondo	700 million tons	In drilling and geothermal engineering, for manufacture of cement and abrasives, as bulk laxative, wound dressing, etc.
Kyanite	Kaduna, Niger	No data	Used in refractory and ceramics production, porcelain, plumbing fixtures, dishware, electrical insulators, abrasives, etc.
Iron-Ore	Enugu, FCT, Kaduna, Kogi, Nasarawa, Zamfara, Niger	Over 3 billion tons	Making of pipes, making of tool and equipment, machineries for processing foods and drugs, infrastructure for transportation
Lithium	Kaduna, Nasarawa, Niger, Zamfara	No data	Used in the manufacture of batteries, in lubricating greases, in fireworks, etc.
Wolframite	Bauchi, Kaduna, Kano, Kwara, Nasarawa, Niger, Zamfara	No data	A source of tungsten used in electrical filaments, also used for making armour-piercing ammunition
Molybdenite	Plateau	No data	For making alloy, stainless steel, in manufacture of armour, aircraft parts, filaments, electrical contacts, etc.
Dolomite	Kogi, Oyo, Edo, Kwara and the F.C.T., Abuja	No data	Ornamental stones, concrete aggregates, a source of manganese oxide, in marine aquariums, to lower acidity in agriculture, etc.

Source: Odeyemi (2016).

acquiring the appropriate mining technologies, Nigeria should be able to turn its huge mineral resources into stupendous wealth.

In order to achieve the essential conditionality discussed above, three other important and tangential factors must be addressed seriously and urgently:

a. Enthronement of peace and security of lives and properties throughout the country
b. Bold and sustained war against corruption
c. Subtle control of population explosion.

I want to appeal to Nigerians to please listen to and learn from the following statement of Martin Luther King:

> The prosperity of a country depends not on the beauty of its public avenues, nor on the strength of its fortifications, nor on the abundance of its public buildings but it consists in the number of its cultivated citizens, its men of education and enlightenment of character.

The challenges of STEM in technological development in Nigeria are majorly owned to the weak industrial policies, discouragement of technological growth by Nigeria's colonial masters and the poorly equipped educational institutions. A number of promising solutions have been suggested to solve these challenges; these solutions include the provision and maintenance of infrastructural facilities, education of the populace on the importance of technological developments, imitating the existing goods in market as well as the review of the land policy in Nigeria, in order to make more land available for farming. The policy orientation should be directed towards making more farm land available to the farmer by organizing the pooling of land. The Land Use Act of the 1999 constitution that vests all land in the government should be reviewed to liberalize land ownership to the benefit of the youths. The provision of agricultural subsidies for fertilizer, farm implements and equipment should be put in place. Adequate measures should be put in place to ensure that the inorganic fertilizers be available to the farmer and its racketeering be discouraged. The use of organic and biofertilizers should also be promoted.

Recommendations:

1. The government should promote the establishment of agro-based industries that are capable of processing Nigeria's agricultural raw materials in a most efficient manner. This will add more value to produce from the farm and promote shelf life. There should be emphasis on the local processing of raw crops for the local industries as well as for export. This will create more employment opportunities, a sustainable market for the products and more income will be generated.
2. There is a need for a significant investment in the efficient processing technologies and packaging of the different agricultural products. Models of on-farm storage facilities and technology should be introduced and made available to the farmers.
3. Concerted efforts should be made towards providing credit support for small-scale farmers who constitute 80% of farm holdings and some medium-scale operators. The lending interest rate should be low and not prohibitory.
4. There is the need to put in place an agricultural tariff regime that would protect Nigeria's agricultural produce from unbridled foreign imports and competition. The policy orientation should be towards minimizing food import through high tariffs on such imported foods, especially those that are produced locally, in order to stimulate their local production.
5. In order for the youths to have entrepreneurial success in the agro-allied business, provision should be made for technical training workshops through the extension services. This would be to x-ray the various business needs and diffuse to them the best technique on how to carry them out, in order to maximize their productivity.
6. There is need for the provision of infrastructure such as good road networks and railways geared towards developing crops required for export.
7. Efforts should be made to alter the movement of people from rural communities to urban areas by extending equal social benefits, establishing rational

schemes for agrarian reforms. This will stimulate the hitherto uneconomical small holdings which is typical of our agricultural sector.

8. The 10 million Nigerian children of school age who are presently not in schools must be compulsorily enrolled in schools through the Universal Basic Education to enable them acquire knowledge and skills as essential pre-requisites for wealth generation. Any defaulting state government that will not put all children of school age in schools should be sanctioned for social and economic sabotage.
9. The various organs of the government should be repositioned in order to adequately regulate the educational standard and quality in the numerous primary and secondary schools, Polytechnics and Universities springing up in the country.
10. National skill acquisition centres should be established in all the geopolitical zones of the country by the Federal Government so that secondary school leavers and university graduates can spend 6–12 months to learn some vocations, trades and skills that are pre-requisites to job creation and wealth generation.
11 Similarly, all state and local governments should establish vocational and skill centres to enable school leavers add value to their knowledge acquisition for wealth generation, while anybody above 21 years of age should be admitted without any educational qualification to such centres.
12. The so-called area boys, touts and school dropouts should be resettled at skill acquisition centres to learn trades, vocations and sporting activities in order to make them wealth-generating and responsible citizens.
13. There is need for building and maintenance of new facilities, and provision of teaching materials for all categories of teachers.
14. The periodic training and adequate remuneration of teachers must be of serious concern to all proprietors of institutions.
15. The industries should be encouraged to support joint workshops, exhibitions, seminars and conferences for cross-fertilization of ideas and expertise in the industries, while specialists from the industries do part time teaching in the Universities and Polytechnics.
16. All the stakeholders, especially the National University Commission, National Board for Technical Education, National Commission for Colleges of Education and various ministries of education should address the various serious and incapacitating lapses in our educational policy objectives.
17. The study of science subjects must be made compulsory in all primary and secondary schools as Nigeria urgently needs to produce a huge pod of scientists, engineers and technologists if we hope to be among the 20 most industrialized nations of the world in the nearest future.
18. Similarly, the study of science and technology must be vigorously pursued in all tertiary institutions.
19. Many professionally and academically competent Nigerians in several European and American countries should be encouraged and induced to come back home to assist the socio-economic and technological development of a great Nigeria.

20. Our leaders should eschew corruption and general indiscipline in order to be able to enthrone virtues of probity, accountability, transparency, hard work, discipline and honesty. The parents, leaders and elders are the mirrors that our youths gaze into daily for their guidance. It is not money but intellectual capital that drives prosperity; hence poverty is driven by and sustained by a lack of intellectual capital. For instance, half of the world's population, i.e. about 3.5 billion people live on an average of $500 a year, while Bill Gates, the new millennium Prince of Technology, earns $500 every second. By controlling technology and taxing computer users, Gates has become richer than each of the 70 poorest nations on earth. Hence, by using his financial might, he has conquered more territories than Genghis Khan, Julius Caesar and Alexander the Great, all put together.
21. Finally, it must be understood once and for all that the educational level of a society and economic growth are closely interrelated. Likewise, it is human knowledge and skill that are instrumental to economic and technological progress. The most valuable of all capital is that invested in human beings (Odeyemi, 2014).

REFERENCES

Ajodo-Adebanjoko, A. & Walter, U. O. (2014). Poverty and the challenges of insecurity to development. *European Scientific Journal*, 10, 361–372.

Amaechi, N. M., Umunnakwe, G. C. & Benson, O. V. (2018). Educators' perception of accreditation processes in university-based library and information science schools in South-East, Nigeria. *Library Philosophy Practice*, 1, 1–18.

Bhosale, S. V., Bhilave, M. & Nadaf, S. (2010). Formulation of fish feed using ingredients from plant sources. *Research Journal of Agricultural Sciences*, 1, 284–287.

Collins, T. (2015). Imitation: a catalyst for innovation and endogenous growth. *Economic Modelling*, 51, 299–307.

Compagnucci, L. & Spigarelli, F. (2018). Fostering cross-sector collaboration to promote innovation in the water sector. *Sustainability*, 10, 4154.

Daud, S., Omotayo, A., Aremu, A. & Omotoso, A. (2018). Rural infrastructure and profitability of food crop production in Oyo state, Nigeria. *Applied Ecology Environmental Research*, 16, 4655–4665.

Fried, G., Chauvel, B., Reynaud, P. & Sache, I. (2017). Decreases in crop production by non-native weeds, pests, and pathogens. *Impact of Biological Invasions on Ecosystem Services*, pp. 83–101. Springer.

Heeb, L., Jenner, E. & Cock, M. J. (2019). Climate-smart pest management: building resilience of farms and landscapes to changing pest threats. *Journal of Pest Science*, 92, 951–969.

Henz, G. P. (2017). Postharvest losses of perishables in Brazil: what do we know so far?. *Horticultura Brasileira*, 35, 6–13.

Isioto, N. N., Nathaniel, N., Philip-Kpae, F., Kpae, F. & Rachael, D. (2017). Roadmap towards Nigeria's technological and industrial independence industrial independence. *CARD International Journal of Engineering and Emerging Scientific Discovery*, 2, 91–101.

Jatto, M. K. (2010) Theme: Actualizing Vision 20/202/20: The Role of Academia. The role of Library and Information Services in Actualizing Vision 20/20/20. The Role of Library and Information Services in actualizing Vision 20/20/20., 2010 Federal College of Education (Technical), Asaba, 1–9.

Jin, L., Duan, K. & Tang, X. (2018). What is the relationship between technological innovation and energy consumption? Empirical analysis based on provincial panel data from China. *Sustainability*, 10, 145.

Kennedy, A. B. & Lim, D. J. (2018). The innovation imperative: technology and US–China rivalry in the twenty-first century. *International Affairs*, 94, 553–572.

Ndubuisi, A. G. (2018). British educational management policies in Nigeria: a historical overview. *Electronic Research Journal of Behavioural Sciences*, 1, 52–61.

Nwangwu, G. (2019). Improving agriculture in Nigeria through public-private partnerships. *African Journal of Business Management*, 13, 256–263.

Odeyemi, O. (1991). Protein energy malnutrition in Africa: Bradyrhizobium to the Rescue? In: A. T. Ba and M. Ndoye (eds), *The Role of Biology in Solving the Food Crisis in Africa. African Biosciences Network*, UNESCO-BREDA, Dakar, pp. 760–762.

Odeyemi, O. (2008). Microbes: unseen agents of economic development. *Inaugural Lecture Series*, 214. o.A.U. Ile-Ife: Obafemi Awolowo University Press Ltd, Ile-Ife, Nigeria.

Odeyemi, O. (2014). *Two Centuries of Oil and Gas.* Universal Academic Services, China.

Odeyemi, O. (2016). *Rhizobium Inoculant (Biofertilizer) Production and Utilization for Improved Legume Yields.* ENPOST Nigeria Limited.

Olaniyan, A. O. & Adetoye, D. (2016). Nigeria and the vision 20: 2020 projection: building castle in the air? *Journal of Education, Society Behavioural Science*, 15, 1–8.

Osabohien, R., Afolabi, A. & Godwin, A. (2018). An econometric analysis of food security and agricultural credit facilities in Nigeria. *The Open Agriculture Journal*, 12, 227–239.

Saidu, A. (2011). Harnessing the entrepreneurial components of home economics education in NCE Curriculum for achieving vision 20: 20: 20. *Multidisciplinary Journal of Research Development,* 17, 46–51.

Salami, C. (2011). Entrepreneurial interventionism and challenges of youth unemployment in Nigeria. *Global Journal of Management Business Research*, 11, 17–24.

Serdyukov, P. (2017). Innovation in education: what works, what doesn't, and what to do about it? *Journal of Research in Innovative Teaching Learning*, 10, 4–33.

Siyanbola, W., Adeyeye, A., Olaopa, O. & Hassan, O. (2016). Science, technology and innovation indicators in policy-making: the Nigerian experience. *Palgrave Communications*, 2, 1–9.

Sobel, R. S. (2008). *Entrepreneurship: The Concise Encyclopedia of Economics.* Carmel, IN, Library of Economics and Liberty.

Steinfioff, D. & Burgess, J. (1993). *Small Business Fundamentals*, New York, McGraw-Hill

Uguru, C., Ajayi, S. & Ogbu, O. C. (2015). Strategies for dealing with low adoption of agricultural innovations: a case study of farmers in Udenu LGA of Enugu State, Nigeria. *Journal of Education Practice*, 6, 7–12.

Uwaifo, V. (2010). Technical education and its challenges in Nigeria in the 21st century. *International NGO Journal*, 5, 040–044.

Uwaifo, V. & Uddin, P. (2009). Technology and development in Nigeria: the missing link. *Journal of Human Ecology*, 28, 107–111.

World Health Organization (WHO) (2015). Health in 2015: from MDGs, millennium development goals to SDGs, sustainable development goals. Geneva, WHO Press. ISBN 978 92 4 156511 0.

11 Capacity Building in Contemporary Biotechnology in Nigeria

History, Impact and Way Forward

Sylvia Uzochukwu
Federal University

Christie Onyia
Godfrey Okoye University

Nwadiuto (Diuto) Esiobu
Applied Biotech International Nigeria Ltd
Florida Atlantic University

Joan Campbell
Frederksberg Hospital

Paul Keese
Office of Gene Technology Regulator

Ivan Ingelbrecht
FAO/IAEA Laboratory on Plant Breeding and Genetics

Alex Ochem
Cape Peninsula University of Technology

CONTENTS

DOI: 10.1201/9781003179177-11

11.1 INTRODUCTION

The 21st century has been referred to as the age of biology, driven by monumental discoveries and innovations in the field of molecular biology, with gene cloning and DNA manipulations at its core (Glover, 2012). In the 1990s, recombinant insulin, first approved for use in 1982 (Sandow et al., 2015) had already taken over diabetes control in the world. Also, in the Americas and Canada, genetically modified (GM) crops had already enjoyed many years of use, starting with the approval of the GM tomato Flavr Savr in 1992 (Aldemita et al., 2015), but most of Africa was yet to get involved. African scientists, who are otherwise talented in their various fields in the life sciences, lagged in the new biology. The lack of competence in basic molecular biology principles and techniques impacted teaching and learning at all levels. As a result, their graduates were poorly equipped for contemporary scientific research at the cutting-edge of modern biotechnology. The scientists however recognized that the situation put the continent at risk of losing the opportunity to join the rest of the world in achieving food security, affordable health care, sustainable environment and other benefits of the life science revolution.

It, therefore, became imperative to seek strategies for building up a critical mass of home-based scientists competent in the new techniques, who could lead the continent's growth in DNA and gene technology. In Nigeria, human capacity building and training became the top priority of all stakeholders, requiring the need for collaborative efforts of home-based scientists, government, the private sector, diaspora and international bodies/institutions, donor agencies and the Biotechnology Society of Nigeria (BSN).

Short, intensive train-the trainers workshops were visualized as the fastest means of raising the urgently needed critical mass of scientists competent in the new molecular biology techniques in Nigeria, and universities, research institutes and the private sector took up the challenge.

11.2 HISTORY OF BIOTECHNOLOGY CAPACITY BUILDING IN NIGERIA

The federal government of Nigeria, in realization of the role biotechnology could play in solving these problems, set up a new outfit, known as the National Agency for Science and Engineering Infrastructure (NASENI), established in 1992 (NASENI, 1992) of which biotechnology programme was part of until 2001, when National Biotechnology Development Agency (NABDA) was created (NABDA, 2001). NABDA's mandate is promotion, coordination, and deployment of cutting-edge research and development, processes and products for the socio-economic well-being of the nation; while her vision is to promote biotechnology activities that positively respond to national aspirations of food security, job/wealth creation, affordable healthcare delivery and sustainable environment.

To accelerate the development and application of biotechnology to national development, the Executive Chairman of NASENI, late Prof. Godian Ezekwe, set up a seven-man committee to advise the government and develop the road map and time line for achieving the set goals, with the coordinating Director of Science Infrastructure, Hon. Prof. R. A. Borofice as chairman and Dr. Christie Oby Onyia, the first Desk Officer for Biotechnology in Nigeria, as Secretary. Based on their recommendations, the first National Conference on Biotechnology was organized and held in Durbar Hotel, Lagos, in 1993, followed by development of vision 2010 for capacity building in biotechnology and development in Nigeria, with the assistance of UNESCO. The main thrust was to train 1,000 first-generation Nigerian biotechnologists in the focussed areas of fermentation technology, tissue and cell culture techniques and marker-assisted/DNA techniques.

Prior to this, assessment of the existing facilities and human resources capacity needs of the country was carried out by NASENI and the Israeli embassy in Nigeria. The exercise was successfully conducted by two Israeli consultants assisted by Nigerian scientists, with the financial support of both governments. With the adoption of their report, Obafemi Awolowo University, Ile-Ife, and Federal Institute for Industrial Research, Oshodi, were designated centres of excellence for fermentation technology; University of Nigeria, Nsukka, Nnamdi Azikiwe University, Awka, and Bioresources Development and Conservation Programme were mandated to develop capacity in tissue and cell culture techniques, while NASENI (Biotechnology Programme) and International Institute for Tropical Agriculture (IITA), Ibadan, were to conduct training workshops and training in deployment of marker-assisted and DNA technologies for crop improvement and food security, in order to achieve the goals set out in Nigeria's vision 2010 for manpower development. However, before the turn of the century, Nigeria had created enough interest in the adoption of advances in new technologies, including space science and biotechnology, resulting in the creation of two agencies known today as: (1) National Space Research and Development

Agency (NASRDA), founded in 1999, and (2) establishment of advanced laboratory for biotechnology research and development in Sheda Science and Technology Complex (SHESTCO), in 1994, and NABDA in Abuja 2001.

In the realization of the importance of public-private partnership (PPP) in this sector, the Nigerian government, under the leadership of President Olusegun Obasanjo, launched the Diaspora Science Day, thereby throwing the door open for collaborations between scientists and entrepreneurs at home and abroad – fostering capacity building and training in different areas of endeavour, including biotechnology.

11.3 EARLY CAPACITY BUILDING EFFORTS IN DNA TECHNOLOGY TECHNIQUES IN NIGERIA

11.3.1 Foundation for African Development through International Biotechnology (FADIB)

FADIB was founded in 1992 by Professor Nduka Okafor, a Professor of Microbiology then at the University of Nigeria Nsukka. He was perhaps the first to make spirited efforts towards developing a critical mass of African scientists to drive the DNA revolution in Nigeria and sub-Saharan Africa. FADIB's annual hands-on training workshops were basically on gene cloning and recombinant DNA technology through sourced grants from the United Nations Development Programme (UNDP), United Nations Environment Programme (UNEP), etc. Resource persons, with relevant expertise, were drawn from across the globe, until 2006, when he stepped aside as the convener. FADIB has however continued with its annual activities, now limited to national conferences and workshops under the leadership of Prof. James C. Ogbonna of the University of Nigeria, Nsukka.

11.3.2 DANIFOL Biotechnology Consult Limited

DANIFOL Biotechnology Consult Limited was founded and registered in 1998 by Dr. Daniel Olukoya, a molecular biologist and then a researcher at National Institute for Medical Research (NIMR), Yaba, Lagos. The outfit, after inception, trained and issued certificates in Molecular Biology Techniques: Genome Analysis for a few years. Dr. Daniel Olukoya was a member of the seven-man committee set up by NASENI to draft the blueprint for development of Nigeria's biotechnology, of which Dr. Christie Onyia was the Secretary.

11.3.3 Contributions of the University of Agriculture Abeokuta (UNAAB, now Federal University of Agriculture, FUNAAB)

The next major sustained impetus for capacity building in modern biotechnology and DNA techniques came from UNAAB, now FUNAAB (Uzochukwu 2008, Uzochukwu, et al., 2012).

11.3.4 The Role of the Nigerians in the Diaspora

Many of the home-based Nigerian scientists who felt ill-prepared for the DNA revolution in their various fields sought to address this deficiency by seeking short training courses abroad, but this approach was both expensive and limited in impact. In 1999, a Nigerian scientist in the Diaspora, Dr. Joan Campbell-Tofte, herself a former undergraduate and Master's student of Prof. Nduka Okafor, and an Edinburgh-trained molecular biologist, rationalized that intensive short training courses organized in Nigeria by a Nigerian expert domiciled in a western country will be the fastest way to address this need and build human and infrastructural capacity in the new DNA technology. She envisaged that lectures and laboratory work spanning from fundamental to applied topics presented to Faculty would allow for maximal learning outcomes. She collaborated with Dr. Sylvia Uzochukwu, then of UNAAB, who assembled a team in UNAAB, and in 2000, they commenced a series of annual 2-week intensive hands-on training courses in DNA manipulating techniques, Bioinformatics and Biosafety of GM organisms. The courses were known as the UNAAB Summer Courses in Biotechnology and were designed to re-train scientists in the life sciences in Nigeria, on techniques for manipulating nucleic acids. Dr. Campbell-Tofte sourced the grants for the first two years, from Denmark, to support her participation and to purchase vital equipment and reagents. She bought the first molecular biology equipment in FUNAAB from these grants and single-handedly taught the practical classes and mentored participants, those first years. The university showed its appreciation, by creating a Biotechnology Centre, in 2001, after the second course. It was probably the first in the country under the Pioneer Directorship of Professor Isola Adamson.

Other Nigerian scientists in the Diaspora worked with the group at different times including Dr. Alex Ochem, a molecular biologist, who supported the work of the group every summer for the training workshops from 2005 to 2009, by his employers, the International Centre for Genetic Engineering and Biotechnology (ICGEB), Trieste, Italy. He helped source grants and reagents, taught the courses and mentored scientists.

Professor Diuto Esiobu, another Nigerian biotechnology expert at Florida Atlantic University, USA, in the Diaspora, member of the Board of the Education Committee of the American Society for Microbiology in Washington DC, was invited by the Obasanjo administration to give the keynote address at the first annual Nigeria Science Diaspora summit 2005 in Abuja. Following a highly insightful talk on the applications of biotechnology in nation building and development, she was introduced to the UNAAB team by NABDA in 2006, to help with their re-training programme. She joined the UNAAB team's summer workshop efforts from 2007 to 2009. Applying the proven active learning, intensive short course approach, already in use by the American Society for Microbiology (ASM), Prof Esiobu introduced transformative pedagogy (Esiobu et al., 1999) helped source grants for her participation, equipment and reagents and mentored scientists on ground. Altogether, a series of 14 training workshops responsible for the paradigm shift in biotechnology manpower development in Nigeria between 2000 and 2009 were designed and presented by the UNAAB group.

11.3.4.1 The Role of the International Scientists in Nigeria

The secret of the success and sustainability of these initial courses was collaboration with Nigerians in the Diaspora, but the secret of the survival of the courses at all lay with the international scientists within the country. The Biotechnology Laboratory of IITA played a major role in nurturing the courses within the first five years, with its scientists serving as resource persons and bringing their equipment. This collaborative effort between the organizing Nigerian scientists, the Nigerian scientists in the diaspora and international scientists in Nigeria was the key to the success of these re-training courses.

11.3.4.2 Target Participants

Target participants included scientists from national research institutes, university lecturers and post-graduate students, and on one occasion, a special one was held just for final year undergraduates in the life sciences.

11.3.4.3 Course Curriculum

Scheme of work included active learning and hands-on exercises, lectures and laboratory practical work covering fundamentals of molecular biology, gene structure, gene cloning, gene expression and analysis, DNA manipulation, as well as extraction and purification of proteins. All participants presented personal projects, and team projects were built into the work schedule. Assessment of learning outcome was based on responses to questionnaires after each course as well as written tests and peer evaluations of team projects and presentations.

The original course curriculum and manual was developed by Dr. Joan Campbell-Tofte, now of the Department of Clinical Biochemistry, Frederiksberg Hospital, Denmark, and was later improved with input from the other scientists in the Diaspora – Prof Diuto Esiobu of Florida Atlantic University, USA, and Dr. Alex Ochem of ICGEB, Trieste, Italy.

11.3.4.4 Faculty

Participating Faculty consisted of the organizing group of Scientists at the University of Agriculture, Abeokuta, the Nigerian Diaspora (scientists) practising in developed countries and International Scientists in Nigeria. The scheme benefited from a Federal government initiative which brought home notable Nigerian scientists in the Diaspora, during the summer, and invited universities to send local scientists to interact with them. It was at such Diaspora meetings that Drs. Ochem and Esiobu were introduced to the UNAAB team.

Guest lecturers were invited to lecture on important related topics, especially Biosafety, Bioinformatics and Intellectual Property Rights (IPRs). Such Guest Lecturers were usually drawn from the Biosafety Desk of the Ministry of the Environment, which was the Focal Point for Biosafety in Nigeria, as well as from NABDA, and SHESTCO.

As already mentioned, international scientists at the nearby IITA were an important reason the FUNAAB re-training workshops survived the first five years. Collaboration with locally available expertise was vital for the success of this enterprise. The authors of this chapter are the international and national resource persons during the period

2000–2009, and the leader of the local team at FUNAAB, Prof. Sylvia Uzochukwu. The names of the other members of the local team are given later in this chapter.

11.3.4.5 Funding

Support came in cash and kind from international donor grants, first at the instance of the Nigerian scientists in the Diaspora and later at the instance of the FUNAAB team. The success of the first two courses created donor-confidence, and the grants grew from very modest few thousand dollars to hundreds of thousands of USD. Counterpart funding from the university consisted of provision of laboratory space, board for visiting experts, salaries of the local personnel and other facilities such as electricity and water. The university responded to the activity of the group not just by creating a biotechnology centre in 2001 but also by following up thereafter, with the erection of a beautiful befitting biotechnology centre building in 2008. The solid support of the university was the most important factor responsible for the take-off and sustenance of the courses (Table 11.1).

TABLE 11.1
Summary of Donor Support for UNAAB Summer Courses in Biotechnology, 2000–2009

S/N	Donor	Grant to	Year	Purpose of Grant
1	World Bank STEP-B Project	Prof. Sylvia Uzochukwu and team	2009	Equipment for the new biotech centre, air fare and full support for two experts, and for one short course
2	International Centre for Genetic Engineering and Biotechnology (ICGEB)	Dr. Alex Ochem and Prof. Sylvia Uzochukwu	2009	Full support for one short course
3	Academy of Science for Developing Countries (TWAS)	Prof. Diuto Esiobu and Prof. Uzochukwu	2008	International expert grant for air fare and support for expert
4	American Society for Microbiology (ASM) USA	Prof. Diuto Esiobu and Prof. Sylvia Uzochukwu	2007	International professorship grant for air fare, equipment and support for expert
5	International Centre for Genetic Engineering and Biotechnology (ICGEB)	Dr. Alex Ochem	2005–2009	Air fare and support for expert, and reagents
6	USAID NARP Project in Nigeria: sub-grant	Prof. Sylvia Uzochukwu	2004	Equipment and tuition support for one short course
7	International Institute for Tropical Agriculture (IITA)	Dr. Paul Keese and Dr. Ivan Ingelbrecht	2000–2005	Loan of equipment, consumables and faculty
8	Society for General Microbiology (UK)	Dr. Campbell-Tofte	2001	Air fare, equipment and reagents, and support for expert
9	Plasmid Foundation, Denmark	Dr. Joan Campbell-Tofte	2000	Air fare, equipment and reagents, and support for expert

11.3.4.6 Course Fees

Whenever there was a moderate grant, the courses were tuition-free so that the younger scientists, who really needed the training, could attend. When there was no grant or the grant was small, the participants had to pay a minimal tuition fee, to cover cost of reagents and logistics. It was not the intention of the organizers to use the course as a source of internally generated revenue for the university. It was not for profit but entirely for service to the country. For all courses, participant selection was based on perceived need, as judged from applicants' curriculum vitae.

It is important that efforts towards similar re-training initiatives be directed towards making them tuition-free, and never seeing them as a source of revenue for hosting institutes and organizations. When tuition is paid in Nigeria for such training courses, most of the participants will be those who don't need the course, such as friends and cronies of chief executives.

11.3.4.7 Impact of Training Courses

Over 500 scientists from 46 tertiary institutions were trained in 14 1- to 2-week intensive courses. Courses were at first for two weeks, but experience showed that 1 week was optimal duration for the course. Longer courses wore out the resource persons due to the high intensity of the courses. While the vast majority of course graduates were Nigerians, a small percentage came from Togo, Argentina, Benin, Ghana, Iran, Cameroon and Rwanda. Post course assessment and research proposals presented by participants at the end of courses showed 40%–80% improvement in grasp of key molecular biology techniques and applications presented. Nearly 100% of participants acknowledged the courses as relevant and eye-opening.

Project impact

- A world-class Biotechnology Centre Building at the Federal University of Agriculture, Abeokuta, built by the university authorities in response to the course activities, and equipped with a World Bank STEP-B grant, also in response to the course activities.
- A much-increased national awareness for modern biotechnology and biosafety was achieved – 500 scientists from 46 tertiary and research institutions.
- Up to 40 tertiary institutions and research institutes in Nigeria established biotechnology centres and departments, often initiated and headed by alumni of the courses, who then began to organize their own training courses.
- Revision of biotechnology curricula in Nigerian universities, and establishment of policy to guide it by the National Universities Commission, in response to the increased biotechnology awareness.
- The condensed intensive courses were relatively inexpensive platforms compared to observed transformational impact created.

The FUNAAB Home Team 2000–2009 (Titles are as at 2009)

- Prof. Sylvia Uzochukwu
- Prof. Mobolaji Bankole
- Dr. Yinka Edema

- Dr. Akin Popoola
- Dr. Emily Ayo-John
- Emmanuel Idehen
- Dr. Toyin Opeolu
- Dr. Andrew Agbon
- Johnson Oyewunmi
- Dr. Tumi Adebambo
- Dr. Martha Bemji
- Dr. Olusegun Atanda
- Dr. Peter Akintokun
- Mr. Ibiwunmi Oloye

11.3.5 Creation and Role of Applied Biotech International Nig. Ltd (ABINL) in Building Biotech Expertise

As the myth of the new DNA science was being decoded and many researchers, academic staff and technologists were being trained, two major gaps and unresolved needs became apparent. First, the urgency of the need to scale the train-the-trainer effort; and second, the lack of any well-established functional laboratories in most of the universities and government research institutions meant that the workshop participants could not grow the new skills learned. Even when funds or donations were available, the paucity of basic understanding and skills made it near impossible for the economy of scale to apply. Many of the participants at the workshops lamented that they produced graduates who were outdated in the fundamentals and applications of contemporary life sciences! To meet the urgency to widen the training of trainers and take the emerging science to home institutions where more persons will have access, Prof. Diuto Esiobu launched the first major Nigerian private biotechnology company – Applied Biotech International Nigeria Ltd (ABINL) in 2006. ABINL rapidly developed mobile training labs with the founders' personal funds, partnered with the National Biotechnology Development Agency, under Professor Bamidele Solomon and other home-based scientists to take the innovative learning techniques to universities and government research institutes, reaching participants from nearly all the six geopolitical regions of the country. Between 2006 and 2016, more than 600 participants from all geopolitical regions of the country had been trained, creating the paradigm shift that created a whole generation of current biotechnology leaders in Nigeria. By taking the training to the universities and research institutes, capacity building workshops became more cost-efficient and allowed local post-graduate students and technologists, who would not have been able to afford the travel and lodging expenses to distant locations, to attend and participate. At the same time, applied biotech staff was able to provide free consultancy for many universities and helped to procure equipment, design laboratories and establish functional world-class biotech centres around the country. ABI Nigeria is run as a social enterprise where the funds realized from setting up labs are ploughed back into developing a world-class laboratory at Wuye, Abuja, for researchers and graduate students to come and receive personalized training and mentoring in advanced biotech research protocols. Sample drop-off services are also available for molecular analysis of all types of samples (microbes, animals, plants and biodiversity). In addition, ABI Nigeria facilitates regular workshops beyond the basics at its headquarters in Abuja, thus complementing the basic molecular biology workshops of UNAAB and other universities. In collaboration with the BSN and international partners, ABINL is working to introduce accreditation and certification exams covering theoretical and laboratory skills to ensure a quality and skilled workforce in Nigeria.

To provide real-life, practical experience in biotechnology for the growing population of young learners, Applied Biotech Nigeria established internship programmes and hopes to expand its reach as resources grow. In addition to providing paid internships to undergraduates in biotech, the privately run company provides research scholarships for graduate researchers at the annual BSN international meetings and helps to organize and run the pre-conference workshops at the widely acclaimed annual international conferences of the BSN. Further to providing quality research materials to laboratories, ABI also assists Nigerian scientists by bringing international partners for networking and expanded services. The full suite molecular biology laboratory is equipped with world-class functional equipment for microbial culture, food, water and soil quality tests, metagenomics, plant and animal barcoding, protein analysis and functional genomics research. With its partners and collaborators, applied biotech guarantees minimal sequencing costs – Sanger and next generation, as well as metagenomics research and analysis.

For more than one decade, Applied Biotech Nigeria has operated with the vision of powering Nigeria's bioeconomy through development of skilled workforce, leadership in research direction and design and equipping of research infrastructure to address all aspects of the sustainable development goals. The company is honoured to be among the leading experts involved in designing and creating the bioeconomy strategic blueprint for Nigeria. Beyond the shores of Nigeria, Applied Biotech Nigeria is helping to shape the Global Bioeconomy through international advocacies and hosting a workshop at the Global Bioeconomy Summit 2020 (www.GBS2020.net). As the needs of Nigeria as a country grows, applied biotech diversifies its modalities, investing into virtual training equipment and expanding its programmes into a department of accredited tertiary molecular diagnostics laboratory for detection and precise identification of human, animal and plant pathogens, including all viruses, bacteria and fungi. The lack of adequate capital, limited policy support (tax rates, power & infrastructure resources) and complicated business ecosystem remain the major handicaps in growing and scaling private biotech companies in Nigeria. Yet, the private sector is pivotal to shaping the economy and job ecosystem through translation of research into products and services. A deliberate path to support a healthy growth and participation of private companies will provide wealth and jobs for the burgeoning brilliant Nigerian youth.

11.3.6 The Role of NABDA in Capacity Building

The initial focus of NABDA was developing capacity in Bioinformatics, as follow-up on the achievements of NASENI and other stakeholder in training in biotechnology. The first NABDA's national workshop on Bioinformatics was held in 2003 under the supervision of Prof C. P. E. Omaliko as the Director General of NABDA, who subsequently created the Department of Bioinformatics and Molecular Biology. Several workshops and trainings were conducted in partnership with National Information Technology Development Agency (NITDA), IITA, etc. NABDA acknowledges the contributions of biotech registered companies, especially Applied Biotech Institute (ABI), Nigeria, headed by Prof N. Esiobu for her concerted efforts in holding several trainings/workshops in collaboration with NABDA. These collaborative efforts with ABI and other individuals both at home and in diaspora led to the development of an award-winning grant proposal – the STEP-B projects. The over $54 million grant, secured under the dynamic leadership of Prof. B. O. Solomon, enabled NABDA to equip her zonal centres located in North-East Zonal Biotech Centre in the University of Maiduguri; South-East Zonal Biotech Centre in the University of Nigeria, Nsukka; South-West Zonal Biotech Centre in the University of Port-Harcourt; and NABDA headquarters in Abuja. The availability of these state-of-the-art equipment has accelerated capacity building and training in Nigeria, thereby enabling the country to leap-frog into the new era of application of biotechnology in national development.

11.3.7 DNA Learning Center Nigeria (DNALC) – Nigeria

- The DNALC-Nigeria was established to pioneer a shift from theoretical instruction of modern biology to hands-on and inquiry-based learning. The centre was conceived through a collaboration between Cold Spring Harbor Laboratory (CSHL), Bowie State University and GOU. It is the first hands-on teaching institution of its kind in Africa. The objective of DNALC-Nigeria is to offer the same hands-on lab experience in Nigeria as is offered at the DNALC in Cold Spring Harbor, providing teachers, professors and students with lab and analytical skills needed for scientific discoveries for the benefit of mankind. We prepare students to pursue successful graduate and post-graduate research in American universities.

DNALC, an operating unit of Cold Spring Harbor Laboratory is the world's first science centre, devoted to public genetics education and research. The DNALC is the world's largest provider of student lab instruction in molecular genetics, operating teaching laboratories throughout Long Island and Manhattan and more recently opened in China and Mexico.

DNALC-Nigeria was established in the permanent site of Godfrey Okoye University, located at Ugwuomu, Nike, Enugu, in 2019, after receiving licence, worth $300,000.00 from DNALC NYC, in support of the centre. The objective of the centre, which was conceived through collaboration between CSHL, Bowie State University and GOU, is to pioneer a shift from theoretical instruction of modern biology to hands-on and inquiry-based learning.

The vision of the centre is to make it a reference point of excellence in scientific training and research, while the mission is to support molecular biology training and research by facilitating adequate hands-on laboratory experience in Nigeria and Africa at large. Total number of teachers and faculties trained at the lab till date is 117. The total number of Alumni of Godfrey Okoye University that have benefited from the study abroad programme is nine as at 2019 and two more have been admitted to benefit.

11.3.7.1 Board of Trustees – DNALC-Nigeria

The board of trustees of DNALC-Nigeria was inaugurated at Godfrey Okoye University on Thursday, 7 January 2021, by the Executive Director, Dolan DNA Learning Center (DNALC), Dr. Dave Micklos, D.Sc.

Memberships include:

i. Prof. Bartho Okolo – Chairman, Board of Trustees, Godfrey Okoye University, Enugu
ii. Rev. Fr. Prof. Christian Anieke – Vice Chancellor, Godfrey Okoye University
iii. Prof. Geoarge Ude – Chair, Faculty Senate, Bowie State University, Maryland, USA
iv. Dr. Dave Micklos – Executive Director, DNALC New York, Cold Spring Harbor Lab, USA
v. Prof. A. M. A. Imevbore, Pro Chancellor and Chairman of Governing Council of Joseph Ayo Babalola University, Osun
vi. Barr (Mrs) Joy Ogbonnaya – Legal Practitioner & President, Pleasant Gathering International
vii. Dr. Christie Onyia – Head of Department, Biological Sciences, Godfrey Okoye University, Enugu
viii. Mr Michael Okolo – Assistant Lecturer, Godfrey Okoye University, Enugu, Secretary to the Board.

11.3.8 Overall Exponential Effect

Alumni of these training courses were encouraged to mount their own courses and train their colleagues and students. This resulted in a proliferation of basic biotech training short courses, biotechnology centres and biotechnology programmes, all over Nigeria, like a chain reaction. The synergistic collaborations and the subsequent cascade of activities were primarily responsible for creating the biotechnology workforce that exists in Nigeria today. This helped to generate a formidable network of biotechnology stakeholders and more private biotech enterprises in the country, with the BSN playing a pivotal professional role.

11.4 CONCLUSIONS AND RECOMMENDATIONS

Competent manpower can be generated in new areas of science in a short time, in a cost-effective manner, using the collaborative approach described herein. African scientists in the Diaspora have an important role to play in initiating and supporting such interventions. Committed and dedicated collaborators at home are a vital requirement for the survival of such schemes. Sincere Institutional support for home-based collaborators is a sine qua non for success.

Some critics view the proliferation of contemporary molecular biology short training courses in Nigeria negatively, pointing out that many of the organizers were themselves mediocre in the field. The earlier narrative in this chapter shows clearly that these organizers do not claim to be experts themselves but only facilitated the meeting of the experts and the novices. They have played the very important role of pointing the country's young scientists in the right direction. One of the objectives the training courses hoped to achieve was to demystify DNA technology and create awareness and a hunger for more, in fertile minds. It is hoped that the yearning will drive young scientists in search of more, and they will come back to Nigeria fully empowered as experts to launch Nigeria into the Age of the New Biology. It is gratifying to note that this is beginning to happen.

REFERENCES

Aldemita, R. R., E Reaño, I. M., Solis, R. O., and Hautea, R. A. (2015) Trends in global approvals of biotech crops (1992–2014). *GM Crops & Food*, 6(3):150–166; DOI: 10.1080/21645698.2015.1056972.

Esiobu, N., Fisher, L., Fulks, J., Kandel, J., and Sherwood, L. (1999) *Laboratory Investigations of Topical Issues in Microbiology*. American Society for Microbiology Education Publications 18pp. https://www.asmscience.org/content/education/curriculum/curriculum.81.

Glover, A. (2012) The 21st century: the age of biology. OECD Forum on Global Biotechnology, Paris 12 November 2012. oecd.org/sti/emerging-tech/A%20Glover.pdf.

Juergen, S., Wolfgang, L., Reinhard, B., and Gerhard, S. (2015) Equivalent recombinant human insulin preparations and their place in therapy. *European Endocrinology*, 11(1):10–16 DOI: 10.17925/EE.2015.11.01.10.

NABDA (2001) *NABDA Policy Document*. https://nabda.gov.ng/about-us/ Accessed: 20/01/2021.

NASENI (1994) *NASENI Policy Document*. https://en.wikipedia.org/wiki/National_Agency_for_Science_and_Engineering_Infrastructure. Accessed: 20/01/2021.

Uzochukwu, S. (2008) Capacity building in biotechnology in Nigeria: the UNAAB experience. *Journal of Biotechnology*, 136:S769.

Uzochukwu, S. V. A, Campbell-Tofte, J., Esiobu, N., Keese P., Inglebrecht, I., and Ochem, A. (2012) Proven strategies for building capacity in biotechnology and biosafety in developing countries: an integrated short course approach. Presented at the 12th International Symposium on the Biosafety of Genetically Modified Organisms, St Louis, Missouri, USA, 16–20 September 2012.

12 Outlook on Bioeconomy in Africa

Strategies for Transforming Biotech Knowledge and Technical Base into Skilled Workforce and Productivity

Nwadiuto (Diuto) Esiobu
Applied Biotech International Nigeria Ltd
Florida Atlantic University

Ivar Virgin
Stockholm Environment Institute

CONTENTS

DOI: 10.1201/9781003179177-12

12.1 INTRODUCTION: BIOECONOMY AND BIOTECHNOLOGY

Historically, the global economy, including the economies in Africa, has been driven by the exploitation of natural resources, frequently leading to resource depletion and in some cases also to serious ecosystem degradation. The dependence on intensive use of fossil oil, coal and gas, especially in the developed and emerging economies, has led to serious environmental challenges and climate impacts. However, in looking to the future, there is an opportunity for countries in Africa to avoid the pitfalls of a fossil energy driven economy by using their bioresources as a strategic base for sustainable economic growth, social and inclusive development. Today the promotion of a bioeconomy is highly placed on the political and business agenda for many countries globally, as a major strategic driver for the transformation of bio-based sectors for sustainable economic growth and development. Bioeconomy growth offers the opportunity for countries in Eastern Africa to achieve many of the Sustainable Development Goals, making use of the region's abundant natural resources, including under-utilized agricultural waste materials, to produce value-added products with applications in many sectors to produce food, energy, and industrial goods, thereby creating jobs, generating wealth, and connecting smallholder farmers to new bio-based value chains. Definitions of bioeconomy are evolving and will continue to change over time. The International Advisory Council on Bioeconomy (2018) defines bioeconomy as "*the production, use and conservation of biological resources, including knowledge, science, technology and innovation related, to provide information, products, processes and services in all economic sectors, in order to move towards a sustainable*". A central feature of the bioeconomy is that *scientific research, knowledge and innovation can be applied not only for the production of food, feed, fibre and fuel but also to produce a wide range of agro-industrial and value-added products.* Another critical element of the bioeconomy is *to build value around local bioresources*, maximizing and using all parts of primary produce and the products thereof. Bioeconomy similarly integrates the use of bioresources – all nucleic acid-derived materials with the complementary tools and principles of chemistry, microbiology, biology, physics, economics, and biotechnology (genetic engineering, synthetic biology, industrial fermentation, etc.) to create new and value-added products, processes, and services in a climate-smart and sustainable way. In other words, bioeconomy, if developed sustainably, is the management and engineering of the biosphere at economic scales to benefit all the inhabitants of the earth and the abiotic planet.

In 2018, the bioeconomy was identified as a policy priority by the G20 Meeting of Agriculture Ministers in Argentina because of its contribution to soil and environmental sustainability, food security, ICT application, and more (G20 Meeting of Agriculture Ministers, 2018).

Modern biosciences and biotechnology are key drivers in the development of bioeconomies worldwide. The multidisciplinary nature of biotechnology makes it difficult to articulate an all-encompassing definition. Indeed, the vastness of the biosphere and the enormous depth of its genetic, physiological, and biochemical processes defy a single construct or description of the subject. The consensus from governments, industry, and academia is that biotechnology is the suite of knowledge and techniques that use living organisms (or portions of living systems) – animals, plants, and microorganisms – for a specific output such as products (Dunham et al., 2002) for improvement of life across the globe. Even this widely used definition fails to include emerging vibrant fields in biotechnology that employ non-living entities such as viruses and viroids to create products.

Africa is endowed with a vast portion of the world's remaining arable land and unparalleled biodiversity (genetic resources) (Senwitz et al., 2016), many of which are vanishing rapidly; but what use is a million acre of arable land when its output pales in comparison to that from one acre under appropriate technology (hydroponics, soil fertility, rhizobiomes inoculation, and modern bioscience-based plant breeding)? Another facet of the rapidly expanding field is synthetic biology and cloning which have been used to create cells, proteins, and products that do not exist in nature, expanding the realms of current biodiversity and bioresources. The technology to clone or copy most of Africa's genetic resources in medicinal herbs, resilient plants, and animal genomes, among others, are readily available. Despite having an abundance of oil and natural reserves, terrestrial and maritime resources, African countries must prioritize knowledge-driven economies powered by wholesome technologies to keep pace with global relevance. Adopting contextual bioeconomy policies and development strategies should be accompanied by a robust commitment to developing strong expertise in biotechnology, which is the intersecting tool for both circular and bioeconomies. This underscores the critical need to consider investment in biotechnology training and education one of the greatest priorities of not only African countries but of any nation intending to thrive in the present future.

12.2 GLOBAL TRENDS IN BIOTECHNOLOGY EDUCATION

The unprecedented disruption in global supply chains caused by the COVID-19 pandemic has revealed vulnerabilities in the relationships between production and consumption world systems. Many countries went scrambling for masks, personal protective equipment (PPE), COVID-19 test kits, ventilators, and therapeutics in 2020, while the queue for the hi-tech mRNA and other vaccines could not be longer. These commodities are products of biotechnology, emphasizing the importance of promoting biotech education, building platforms for innovation, and sharing best practices to support the development of sustainable bioeconomy in the health and security of all sovereign nations.

The biotechnology field has undergone very rapid evolution in multisectoral directions. Since the discovery of recombinant DNA in the early 1970s and the subsequent landmark discovery of thermal cyclers for polymerase chain reaction (PCR) (mid-1980s), next-generation sequencing platforms, metagenomic protocols, synthetic biology, and Clustered Regularly Interspaced Short Palindromic Repeats (CRISPR) technologies, among others, the contemporary precision and unlimited horizons in biotechnology have been fuelled by advances in computing power and bioinformatics.

This convergence of the digital and emerging tools makes biotechnology and the life sciences one of the greatest revolutions of the 21st century. As databases of all forms of bioresources – genes, genomes, proteins, RNAs, microbiomes, plants, and animals enlarged, and tools to study biological entities advanced, the insight into the workings and interactions of the biosphere deepened, driving curriculum and refining research and development focus. For example, it has been known that microbes helped to shape and sustain planet earth through a myriad of biogeochemical reactions, but the recent discovery of their invaluable roles in regulating ecosystem function and the wellbeing of all living things through molecular signalling is transforming curricula in the life sciences. Many countries are continually working toward formulating an overall subject scope for the field of biotechnology. One of the most inclusive knowledge areas established in the taxonometric structure for biotechnology are foundations of biotechnology, environment, agriculture, bioprocessing, genetic engineering, biochemistry, medicine, and bioethics (Dunham et al., 2002). Figure 12.1 displays components of a broad-based biotech curriculum that could be adapted in content and scope to suit the specific needs of the region/countries.

In the United States of America, biotechnology concepts are introduced at the primary to secondary school level, (K-12) where middle and high schoolers (junior and senior secondary schools) in Biology must show competence in the central dogma of life and gene expression prior to graduation. To fast-track the development of appropriate and adequate workforce, the United States has a well-developed Biotechnology High School magnate and vocational programmes all over the country. These students are exposed to the basic and highly advanced vocational skills in biotechnology during secondary education and or in vocational institutes such as in Monmouth County Vocational School Districts http://www.bths.mcvsd.org/. The rigour and engagement of secondary schools in biotech (https://pblh.palmbeachschools.org/school_info/five_carat_choice_program/Biotechnology), for example, are deeper than in many postgraduate biotech programmes in Africa.

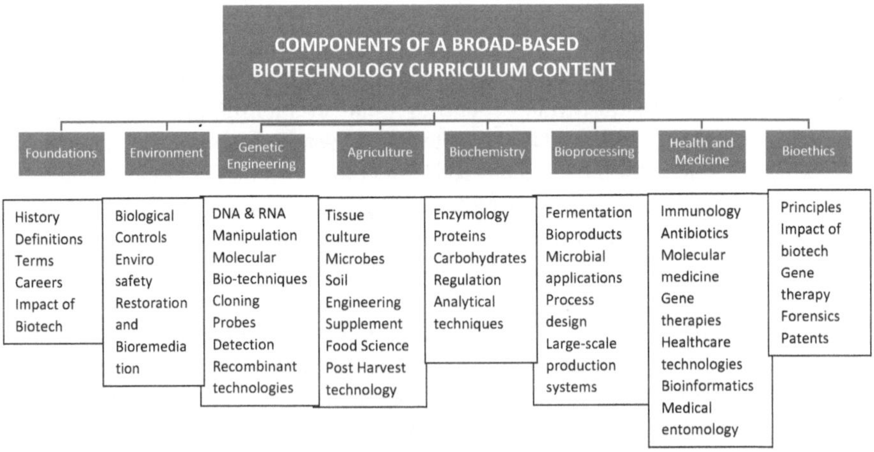

FIGURE 12.1 Broad concepts in biotechnology presented under units for bioeconomy workforce training.

Often, these high schools and their well-trained teachers affiliate with neighbouring universities to boost their students' capacity and research exposure. In addition to the Kindergarten to 12-grade plan, more than 110 universities offer a biotechnology major (https://blog.collegevine.com/list-of-all-u-s-colleges-with-a-biotechnology-major) while numerous others integrate biotech into Biological Engineering, Biology or Life Science Departments. The biotechnology revolution in India started with a strategic focus on training human resources in vaccine development and biopharmaceuticals. The second most populated nation on earth is rightfully tackling youth unemployment through dedicated biotech training and skills development in line with the vast prospects of bioeconomy (https://www.biospace.com/article/biotech-youth-training-program-to-be-offered-in-india/) In Europe, despite the lack lustre enthusiasm for genetically modified organisms, the continent foresaw the huge prospects of the field in building future economies and integrated biotechnology into secondary school curricula in the early 1990s. This gave rise to the European Initiative for Biotechnology Education (EIBE) funded by the European Commission (Harms, 2002). The EIBE project was developed by a multidisciplinary team of scientists and educators and was designed to inform young school pupils and the public on biotechnology. The teaching units included microbes and molecules, fermentation technology, DNA fingerprinting, novel foods, transgenic plants and animals, and environmental biotech, among others.

In fact, four of the seven most populous countries (China, India, United States, and Brazil) on earth have developed clear bioeconomy strategies and are focused on broadening manpower development in biotechnology as the requisite tool. It is critical that the African Union integrates biotechnology in its broadest sense to early education as an investment to its future. Resources and infrastructure in Africa as a whole and Nigeria, the sixth most populous nation globally, must urgently pivot towards growing concrete vocational skills in the science of the current anticipated future and in the context of modern biosciences and biotechnology knowledge.

The booming bioeconomy in countries like the United States of America (as well as India, China, and Brazil) is a direct outcome of investments that deepened the knowledge base of their scientists and human resources. Innovation is the fruit of understanding, just as insight into science and societal needs pave the way for discoveries. Therefore, the necessity for a dedicated commitment to teaching biotechnology concepts and skills at all levels of learning cannot be overemphasized.

12.3 CREATING CONTEXTUALLY RELEVANT CURRICULA FOR TRAINING

The most critical first step to creating a competent and productive workforce is to develop a standard core curriculum. A well-composed curriculum will then guide the employment schedules and laboratory infrastructure development. The curriculum should not be skewed to fit the existing labour force; instead, governments and stakeholders should support the trainers and academic faculty to conform to the standard curriculum through sponsored workshops and incentives. Countries with developed biotechnology enterprises involve multiple stakeholders – academia (all related disciplines), industry, government, and the public in the process of curriculum development.

A good biotechnology curriculum should have a clear goal beyond rhetoric head knowledge and academic illumination. It should seek to deepen learning, ignite curiosity, and empower the trainers and trainees to understand the biotic and abiotic worlds and their interactions. The need to invest in an expansive and detailed/inclusive process could not be overstated, mostly because people do not know what they do not know. Furthermore, every country has its unique suite of economic sectors and industries – biorefineries, agro-industrial, biopharmaceuticals, etc. Conducting and Integrating the Strength, Weaknesses, Opportunities and Threat (SWOT) analysis of the country's or region's biological enterprise and resources is essential in developing a contextually relevant curriculum. It is important to revisit and review such core learning guidelines periodically (5–7 years) following inclusive and expansive consultations with industry, policy, and academia, to keep pace with fast changes, transformations, and technologies.

While it is unrealistic for any single curriculum to cover every important subject, a minimum core expectation baseline should be uniformly adopted, to ensure quality learning of fundamental concepts and skills. A reasonable distribution of biotech training goals should cover both theoretical and practical skills training as well as a solid foundation in the bedrock sciences. Under *Theoretical knowledge*, General Biotechnology (30%), Laboratory Skills/Applications (20%), Chemistry/Biochemistry (14%), Biological Systems (15%), Biological Engineering (10%), and Research Methods (11%) should provide a reasonable weighting of content. For *Practical Knowledge and proficiencies*, the distribution of courses and learning objectives should span Biotech Skills (30%), Basic and Applied Mathematics (20%), Laboratory Devices and Equipment (15%), Laboratory and Biosafety (15%), Databases and Web-based Practicals (10%), and Problem-Solving and Industrial Training (10%). Some of the best practices employ topical practical issues to teach the principles and tools of the science of biotechnology (Esiobu et al., 1999). The emphasis at the undergraduate level should be on building understanding of concrete scientific concepts. Innovation in applied biotech fields is driven by deep insight into the mechanisms of science, rather than through black-box approaches to problem solving. The curriculum for vocational training is skill-intensive with just the adequate amount of background information.

It is not uncommon to find curricula in African Universities with broad-brushed subjects without explicit content and details. Worse still, the materials presented sometimes differ significantly from the topic listed on the curriculum and syllabus, largely because of limited capacity of faculty in the subject area. Although there has been a consistent improvement in the level of literacy of biotech teachers in Africa since 2005 (driven by the hard work of the university faculties, diaspora partnerships, and some government support), a lot of work to train these ambitious and often brilliant African scholars to go beyond the basics remain. In the following sections, some strategies based on the authors' experiences and standard training resources (https://biotility.research.ufl.edu, Bio-Rad's Brown (2018) manual) are presented to provide governments, policy makers, National University Commissions (NUC), academic and vocational institutes, and all stakeholders a framework that can be used to build contextualized biotechnology core curriculum and content. This will in turn help to drive the bioeconomies of the regions.

In Table 12.1, a consolidated detailed content of an example of Masters, undergraduate biotechnology or vocational biotech's knowledge base and practical skill

TABLE 12.1
Theoretical and Practical Competencies Expected of Undergraduate and Graduate Degree Holders in Biotechnology

Theoretical Competency

- **Biochemistry and chemical principles**
  - Types of molecules and chemical bonds
  - Chemistry of macromolecules of life
  - Mixtures and solutions
  - Bases for aerobic and anaerobic reactions/respiration
  - Biological enzymes and catalysts
  - Thermodynamics of cellular metabolism
  - Structure and function of the DNA
  - RNA, transcription and regulation
  - Proteins and the process of translation
- **Biological systems**
  - Cell theory and types of cells
  - Reproduction, mitosis, meiosis, cell division
  - General physiology of cells, including respiration types, photosynthesis, methanogenesis, Nitrogen fixation, xenobiotic degradation pathways
  - Ecology and interaction of cells, including molecular signalling between cells in the environment
  - Ultrastructure of the domains – bacteria, archaea, and eukarya, including types of staining (gram staining, etc.)
  - Genetics of cells and central dogma of life
  - Basic understanding of immunology
- **Fundamentals of biotechnology**
  - Historical advances in biotechnology from discovery of transforming principle (DNA) and career opportunities in the field
  - Techniques used in biotechnology (PCR, types of electrophoresis, blotting, probes and microscopy, sequencing, protein and amino acid sequencing, cell and tissue culture for plants and animals) and their applications
  - Genetic engineering fundamentals and principles
  - Introduction to bioinformatics and synthetic biology
  - Key advances and benefits of biotechnology to the society, including future prospects
  - Biosafety and laboratory safety measures and interventions
  - Ethical considerations in DNA manipulations
- **Theory of practicals in biotechnology**
  - Culture of microorganisms and tissues (plants and animals)
  - Aseptic techniques, sterilization, decontamination, and disinfection protocols
  - Principles of microscopy and types of microscopes
  - Concept of buffers, pH measurements, and metres
  - Nucleic acid isolation and purification, PCR, primer design, and in silico gene manipulation
  - Recombinant DNA technology, transfection, transformation, and shot gun approaches
  - Gene expression, protein and amino acids production. Isolation, purification and sequencing
  - Laboratory equipment used in biotechnology – calibration, maintenance, and validation
  - Sequencing techniques – Sanger, next gen, shot gun, etc. and metagenomics
  - Overview of experimental design, hypothesis development, and the scientific method
  - Understanding research; manuscript and science communication

(*Continued*)

TABLE 12.1 (*Continued*)
Theoretical and Practical Competencies Expected of Undergraduate and Graduate Degree Holders in Biotechnology

Practical & Vocational Skill Proficiency

- **Applied mathematics in biotechnology**
  - Calculations in serial dilutions, solution preparations, molarity, dilution factors and ratios
  - Creation of standard curves and plots plus data interpretation
- **Biotechnology skills**
  - Micropippeting skills and accurate weight measurements
  - Culture techniques, electrophoresis, media and buffer preparations, and asepsis
- **Biotech equipment maintenance and calibration**
  - Centrifuges, weigh balances, spectrophotometers, autoclaves, shakers, incubators, and vortexes
  - Documentation and handling of biological and hazardous wastes, PPE biosafety levels

Applications in Old and Emerging Fields

- Agriculture/food and nutrition – synthetic biology and foods of the future. Soil quality and climate-smart applications
- Health and medicine – proactive surveillance of vectors and viruses, gene therapies, microbiomes and synbiotics, vaccinomics
- Bio-manufacturing & biorefineries – bio-fuels, biomass and bio-plastics, biopharma, bio-textiles and fashion, microbial dyes
- Environmental technologies – bio-based chemicals and products for waste and drinking water treatment, remediation, adaptation technologies
- Technologies for energies – microfactories
- Other commodities of biological origin – cosmetics and fragrance, mushrooms, and leather
- The environment and climate-smart technologies

Graduate education emphasizes applications and problem solving where candidates already possess the requisite bedrock scientific knowledge. Vocational training should emphasize laboratory skills and lab management.

guideline is displayed. These competencies and learning outcomes could be adjusted and weighted by respective institutions based on their mandate and projected manpower needs specified by the bioeconomy strategic blue-print.

12.4 INNOVATIVE APPROACHES TO RAPIDLY CREATE SKILLED WORKFORCE FOR THE BIOECONOMY

Central to the development of a skilled workforce is a competent faculty and teacher community. To avert perpetuation of mediocrity and power the growth of the bioeconomy, the first focus should be training. Teachers are the bedrock of all forms of civilization, spawning both the professionals, knowledge creators and inventors. Indeed, it will not be incorrect to assert that university teachers and instructors are the custodians of a nation's future and economy, being primarily responsible for short- to long-term knowledge base. Empowering such a critical part of a nation's population though seem obvious but is sadly not the case now in many African countries. This negligence must change for the region to participate in the booming knowledge-driven economy. Competitive salaries and conditions of service tied to continuous education and skill would help attract brilliant minds into teaching at all levels of education.

The other half of the equation is how to effectively grow a well-equipped and empowered workforce to run the bioeconomy enterprises. Suffice to mention that biotechnology training is only a starting point towards entrepreneurship in biology. The core curriculum should create an adequate foundation in fundamental principles as well as emerging bioscience concepts and applications. The key to producing requisite workforce that will drive the economies of the 21st century lies at the nexus of foresighted policy, training teachers beyond the basics, vocational skill acquisition, and a deliberate investment into the private sector. These will positively impact transformation of bioresources into value-added products and promote creativity in generating new ones. The dearth of expertise in entrepreneurial biology requires a multifaceted approach at all levels of education to mitigate. Professional organizations, ministries of education, and university systems are working strategically with regional and international partners, stakeholders, and philanthropists, to develop a continental initiative on "Training for the Bioeconomy". Similar to the European initiative on biotech training launched in 1996, such a programme in Africa should be designed and implemented as a multipronged training intervention and bio-based business incubation hubs to reach all levels. Such a programme will simultaneously train all segments of the society to become productive in the bioeconomy.

12.4.1 Inspire Students from Kindergarten to Elementary and Junior/Senior Secondary Schools

There is no substitute to catching them young! Age-appropriate curriculum, including hands-on learning and incentivized Holiday Science Camps, such as the one run by the Applied Biotech Intl Nig Ltd and other entities should be supported and promoted by governments. These examples may be viewed through the following links: https://www.facebook.com/Appliedbiotechng/photos/pcb.603123690047921/603123616714595/; https://www.firstdiscoverers.co.uk/early-years-science-activities-eyfs/. Giving young people opportunities for scientific discovery at an early age fosters the love for science, trains them in scientific concepts, and supports the development of other skills that pertain to innovation (Tanner, 2013; Orhan and Sahin, 2018). The lofty goals and priority agenda 2063 of the African Union enunciates several expected milestones; however, the emphasis on transformative education for the 21st-century science on the continent is missing. Just as no castle could stand without a solid foundation, no bioeconomy or industrialized status can materialize without a robust knowledge of science. Luckily, introducing the concepts of science and of biotechnology in particular to school age children is very inexpensive and can be accommodated by the budget of nearly every school. The first important step is the realization of the need and the campaign to promote science in a broader and inclusive way. Several resources for age-appropriate curriculum and materials are free on the web. Worthy of mention is the well-curated MIT Blossoms Initiative in Math and Science (https://blossoms.mit.edu/mit_blossoms_initiative_math_science_video_lessons_high_school_students). that actively engages the minds of secondary school age children and their teachers. The videos are detailed enough to train the teachers in fundamental principles while allowing students to equally learn alongside.

In addition to Summer Science Summits organized by schools and private sectors and supported by government resources, it is essential to consider creating a systemic structure for delving deeper into STEM by designating some federal or publicly owned secondary schools as magnet high schools in bioeconomy and biotechnology to enable the teachers and students grow the skills of problem-solving. This approach will generate a pool of future workforce in the dominant biology economy of the 21st century. How to assimilate the numerous jobless graduates, youth, and women into high wage-paying jobs that empowers rural and urban Africa alike is the daunting task of the immediate post-COVID era.

12.4.2 Rapid Skill Development and Strategies to Jumpstart the Bioeconomy across Africa

12.4.2.1 Establish Regional Vocational Institutes under Strategic Partnerships to Train Youths for Bio-Jobs

Vocational and technical institutes (VTI) are postsecondary education focused on training students for a specific career or job using hands-on and problem-solving approach. The period of training varies with the intended vocation, ranging from 6 months to 2 years. Its curriculum is centred on the career (nursing, biotechnician, lab technician, biomedical research assistant, electrician, equipment maintenance, R&D and bio-manufacturing, auto-mechanic, cosmetology, chef and culinary experts, etc.). The initiative by the Nigerian government to establish vocational institutes in all six geo-political regions is a step in the right direction. However, the national technical education board should partner with the private sector and other stakeholders to develop a world-class curriculum that uses best practices to implement the vocational training programmes. It is said that creativity is not taught, but it is also true that it can be nurtured. Learning by doing to inculcate the skill of analytical rigour and critical thinking should be the goal of VTIs across Africa. Mediocrity must be avoided by using competent consultant and experts to design and implement these VTIs which can double as career centres for STEM-based jobs.

12.4.2.2 Introduce Short-Term Workshops for Advanced Skill Acquisition and Bioeconomy Concepts

Many pioneer workshops have birthed numerous others across the continent. However, the time is now, for a new type of workshops that go beyond basics or purely academic exercises to programmes with practical linkages to the bioeconomy concepts and Bio-entrepreneurial emphasis. Cultivating the creativity and problem-solving skills of trainees should be the focus of the new breed of workshops. Well-organized short-term workshops employ carefully designed modules with clearly defined learning outcomes (Orhan and Sahin, 2018). For nearly 15 years, various players – private sector, governments, professional societies, funding agencies – have spearheaded development of workshop templates in partnership with home-based scientists and many academic staff to train instructors of Life Scientists and allied researchers. This approach (which is now widely adopted in Nigeria today) engages participants in active-learning strategies, providing opportunities to think and learn the basic principles of DNA manipulations. While these efforts have been widely successful in exposing trainers and postgraduate students to the basics of biotechnology (Uzochukwu et al., 2012), they are hardly sufficient for creating the necessary workforce for the bioeconomy.

Participants at these advanced workshops should be carefully and competitively selected and funded by the government with expectations of a business plan from each participant. Examples of such advanced workshops include bio-manufacturing, skills in biomedical diagnostics, food security and postharvest technologies, biomass transformation technologies, scaling plant tissue culture to business, and quality control in industrial processes. Priority should be postgraduate students, professors, and citizens intending to change their careers. In addition to the technologies, critical thinking skills and entrepreneurial skills should be emphasized by cultivating divergent biological thinking using open-ended questions (Bloom et al., 1956) that leads to problem-solving and innovation. Post-workshop monitoring and evaluation of milestones must be conducted for accountability. Follow-up conferences (could be virtual) of trainee cohorts are international best practices which allow for continued peer support and mentoring.

12.4.2.3 Deploy Mobile Laboratories and or Lab in a Box to Remote Parts of the Country

To reach every corner of the nations of Africa with the urgency of STEM education would require mobile labs or lab in a box. Such laboratories are fully housed within or transported by a vehicle, allowing a highly efficient use of funds in developing and maintaining hands-on practical courses. The mobile labs could be deployed at the local government levels and managed by private partnerships. Every secondary school and even elementary school could easily be scheduled for STEM sessions. It is highly likely that there are many Einsteins and Marie Curies (great scientists of their time) who are wasting in village schools with no science teachers or labs. Adopting the mobile lab approach helps to solve the limited infrastructure and human capital availability in developing countries. Indeed integrating the activities and schedules of such mobile labs with school curriculum for the West African Examination Council or other standardized learning outcomes for Junior Secondary School (JSS) and Senior Secondary School (SSS) will be needed to promote its adoption.

12.4.2.4 Establish Innovation Bio-hubs and Clusters/Platforms

Specific facilities that constitute a 'decentralized platform' could be spread across countries by commodity and/or thematic areas, following the model that was piloted under the Eastern Africa Agricultural Productivity Project (EAAPP), funded by the World Bank. But the new platforms would need to be designed to drive '*research into use*' where multi- and inter-disciplinary links are fostered as well.

Links with NARIs and universities will be a major means of ensuring continuing relevance of the science being taught and its application. In this connection, regional initiatives such as the flagship TAGDev programme of RUFORUM – which *seeks to transform African agricultural universities and their graduates, in order to respond better to developmental challenges through the enhanced application of science, technology, business and innovation for rural agricultural transformation* – will be important to implement partners. Working with universities to develop and/or review graduate programmes to ensure continuing context relevance that reflects both the developments in the science and the evolving needs in the region will be important in this regard. These learning clusters afford the populations wide ranging benefits, including but not limited to the following:

Propelling new partnerships and collaborators: Leverage economic blocks and trans-African trade, Optimize biomass and energy use, Catalyse knowledge sharing, and Mobilize local stakeholders.

12.4.2.5 Leverage Diaspora and International Collaborations

The roles of these key stakeholders in the development of African nations are widely acknowledged. Workforce development and bioeconomy strides are not exempt. How the Nigerian diaspora collaborated with home-based scientists to retrain the first batch of biotech literate faculty have been elaborately discussed elsewhere in this volume. Suffice to say that diaspora were responsible for the bloom of biotechnology and bioeconomy in China, India, and Brazil as well. Invariably the multi-stake partners must work altruistically to train the workforce and help craft the bioeconomy strategic plan in African countries.

12.4.2.6 Online Resources and Virtual Labs and Lectures

Use of the abundant online resources should lead to a quantum leap and transformation in the bioeconomy realm and the STEM education in general. Of course this recommendation assumes that African countries will invest in affordable and constant power supply as well as broadband high-speed internet infrastructure. It is beyond fathom that these crippling obstacles still exist, even in the wealthiest nations of Africa. Private–public partnerships to promote biorefineries, bio-fuels, and bio-energy technologies should contribute significantly in achieving these goals sustainably.

12.4.2.7 Set Up Regional Hubs and Clearing Houses

Whereas the innovation hubs are established by individual countries, districts, or local governments to address the needs of rural and urban populations/add value to in-country biological assets, regional clearing houses are owned by multiple countries to synergistically create bio-jobs and bio-wealth across their borders. The innovation hub or Bio-Innovate of the East African Community (EAC) informs the benefits of such entities. In a broader sense, the African Development Bank and associated agencies could play host or support the strategic development of Africa's bioeconomy and the closely associated circular economies by supporting funding for this critical endeavour. The benefits are extensive, including but not limited to:

i. **Knowledge sharing:** Treasures of indigenous intelligence on survival and sustainability are in silos across Africa. Regional bio-hubs in the economic blocks should facilitate sharing and translating these resources into job-creating wealth within Africa and between Africa and the rest of the world. On the other spectrum of things, the omics technology is rapidly generating rapid growth in the manufacture and adoption of new bio-products around the world. There is an exponential growth in bio-databases and bio-banks of genetic and natural resources in many regions of the world. Yet Africa has zero such centres. As a result, its genetic resources and biodiversity are scattered around the world. These resources, when powered by the next-generation sequencing (NGS) and the bioinformatics pipelines will become enormous assets the continent cannot afford to ignore. In this Century of Biology, it is critical that the African Development Bank and the New Partnership for African Development (NEPAD) prioritize conserving Africa's harvest by consolidating the genomes and phenomes of its human and biological assets into an accessible database that will help to drive economic growth. Regional hubs and clearing houses can fill this gap.

ii. **Efficiency in use of technical expertise:** The dearth of adequate expertise in STEM and Bio-enterprise development in African countries is largely responsible for the slow growth of this transformational sector. Setting up regional hubs and clearing houses will allow regions to collaborate and share human capital and mentors.

iii. **Economic hub and market access:** Multisectoral economic or industrial park would serve as incubators for emerging regional bioeconomy outfits and help to source markets for bio-products. Such economic hubs maintain inventory and trends of bioeconomy as well as other sectors and help to guide international trades and policy within the economic blocks.

12.4.2.8 Accreditation and Quality

When accreditation and licencing is required in the practice of any career, it helps to guarantee premium quality of the players and also ensure that they remain current and up to date in their fields. It creates access to high wage-paying jobs in related employment and industry. A good example is the US Biotechnician Assistant Credentialing Exam (BACE, administered by Biotility). BACE is an industry-recognized exam designed to assess core skills and knowledge sets identified by industry, the academic and performance standards of biotechnology programmes. The exam was originally scrutinized by Bio-Florida, an organization of biotech and related industries in the state of Florida, representing over 3,000 research organizations and companies in the biotechnology, biopharmaceuticals, biomedical devices, and bio-agriculture sectors. The Biotility initiative was conceived under the Workforce Development Program of the Clinical and Translational Science Institute in Florida. All the 21 states in the United States where biotech and bioeconomy industries are flourishing use the services of BACE to accredit their entry level workforce. Individual countries and economic blocks in Africa will do well to develop industry-led and context-relevant curriculum/evaluation to improve quality and output of its workers. In fact, it is highly recommended that more than one accrediting entity be vetted and endorsed by stakeholders as in the accounting field where the Institute of Chartered Accountants of Nigeria (ICAN), for example operates alongside the international Association of Chartered Certified Accountants (ACCA). Such an approach allows for broader access by trainees and for a healthy competition to act as a check on the quality and operations of the accrediting entities.

12.5 CONCLUSIONS AND RECOMMENDATIONS

The need to expand training workshops to go beyond basics is overdue. To develop a competent workforce in biotech and the bioeconomy all trainers of the subject should be accredited and required to take refresher courses periodically. Standard training curricula should be developed with input from multiple sectors, stakeholders, and professional organizations. The formats for training delivery should be diversified to include mobile labs and vocational institutes in order to mobilize brilliant youths who are denied opportunities for lack of access to basic facilities and trainers. It is recommended that a Bioeconomy Council be established under the presidency to coordinate activities in the sector and ensure efficient delivery of economy of scale and quality monitoring as Africa emerges as a major world player. Finally, regional hubs in the economic blocks – Economic Community of West African States (ECOWAS), East African Community (EAC), Arab Maghreb Union (AMU/UMA), and the Southern African Development Community (SADC) – are highly recommended to promote collaborations across borders and promote knowledge sharing within the economic block and between the Africans and their international partners.

REFERENCES

Monmouth County Vocational School District - Biotechnology High Schools, Monmouth County Vocational District, New Jersey . http://www.bths.mcvsd.org/. Accessed online, November 2020.

Bloom, B. S, Englehart, M. D., Furst, E. J., Hill, W. H., Krathwohl, D. R. (1956). *A Taxonomy of Educational Objectives: Handbook 1: Cognitive Domain*, New York: McKay.

Brown, J. K. (2018). *Biotechnology: A Laboratory Skills Course; The Student Edition.* 2nd edition. 200 pp. ISBN 978-0-9832396-2-8.

Dunham, T., Wells, J., White, K. (2002). Biotechnology Education: A Multiple Instructional Strategies Approach. *Journal of Technology Education* 14: 46–81.

Esiobu, N., Fisher, L., Fulks, J., Kandel, J., Sherwood, L. (1999). Laboratory Investigations of Topical Issues in Microbiology. Washington, DC: American Society for Microbiology Education Publications. 18 pp. Available at: https://www.asmscience.org/content/education/curriculum/curriculum.84

G20 Meeting of Agriculture Ministers. (2018). Declaration. Available at: http://www.g20.utoronto.ca/2018/2018-07-28-agriculture.html.

Harms, U. (2002). Biotechnology Education in Schools. Issues in Biotechnology Teaching. *Electronic Journal of Biotechnology* 5 (3): (2002): 5–6. http://www.ejbiotechnology.info/content/vol5/issue3/teaching/01.

International Advisory Council on Bioeconomy – IACGB. (2018) The Global Bioeconomy Policy Report (IV): A decade of bioeconomy policy development around the world. Publisher; Secretariat of Bioeconomy Summit, Berlin; 169 pp. https://gbs2020.net/wp-content/uploads/2021/04/GBS-2020_Global-Bioeconomy-Policy-Report_IV_web-2.pdf

Orhan, T. Y., Sahin, N. (2018). The Impact of Innovative Teaching Approaches on Biotechnology Knowledge and Laboratory Experiences of Science Teachers. *Education Sciences* 8: 2–24.

Senwitz, C., Kempe, A., Neinhuis, C., Mandombe, J. L., Branquima, M. F., Lautenschläger, T. (2016). Almost Forgotten Resources - Biomechanical Properties of Traditionally Used Bast Fibers from Northern Angola. *BioResources* 11 (3): 7595–7607.

Tanner, K. D. (2013). Structure Matters: Twenty-One Teaching Strategies to Promote Student Engagement and Cultivate Classroom Equity. *Approaches to Biology Teaching and Learning* 12: 322–331.

Uzochukwu, S. V. A, Campbell-Tofte, J., Esiobu, N., Keese, P., Inglebrecht, I., Ochem, A. (2012). Proven Strategies for Building Capacity in Biotechnology and Biosafety in Developing Countries: An Integrated Short Course Approach. Presented at the 12th International Symposium on the Biosafety of Genetically Modified Organisms, St Louis, Missouri, USA, 16–20 September 2012.

Index

Note: **Bold** page numbers refer to tables and *italic* page numbers refer to figures.